## The Cambridge Star Atlas

*The Cambridge Star Atlas* covers the entire sky, both northern and southern latitudes, in an attractive format that is suitable for beginners and experienced astronomical observers. It contains maps of the Moon, a series of seasonal sky maps and a small eight-page atlas showing all of the Messier objects, which also serves as an index to the main Atlas charts; a detailed atlas of the whole sky, arranged in 20 overlapping full-color charts. The charts show stars down to magnitude 6.5, together with about 900 non-stellar objects, such as clusters and galaxies, which can be seen with binoculars or a small telescope. Information about these objects can be found in the tables that accompany the charts.

WIL TIRION is the world's foremost designer of astronomical maps. For this fourth edition he has devised improved versions of all the charts, and the text and star data have been completely revised based on the latest information. Clear, authoritative and easy to use, *The Cambridge Star Atlas* is an ideal reference atlas for sky watchers everywhere.

W9-BDM-791

# THE CAMBRIDGE STAR ATLAS

FOURTH EDITION

WIL TIRION

3 1489 00615 2753

$32.99

CAMBRIDGE UNIVERSITY PRESS
Cambridge, New York, Melbourne, Madrid, Cape Town, Singapore, São Paulo, Delhi, Dubai, Tokyo, Mexico City

Cambridge University Press
The Edinburgh Building, Cambridge CB2 8RU, UK

Published in the United States of America by
Cambridge University Press, New York

www.cambridge.org
Information on this title: www.cambridge/9780521173636

© Cambridge University Press 1991, 1996
© Wil Tirion 2001, 2011

This publication is in copyright. Subject to statutory exception and to the provisions of relevant collective licensing agreements, no reproduction of any part may take place without the written permission of Cambridge University Press.

First published 1991
Second edition 1996
Reprinted 1998, 1999
Third edition 2001
Reprinted 2004
Reprinted with correction 2005
Reprinted 2007
Fourth edition 2011

Printed in the United Kingdom at the University Press, Cambridge

*A catalog record for this publication is available from the British Library*

*Library of Congress Cataloging in Publication data*

Tirion, Wil, 1943–
Cambridge star atlas / Wil Tirion. – 4th ed.
p. cm.
Includes bibliographical references.
ISBN 978-0-521-17363-6
1. Stars – Atlases. I. Title.
QB65.T537 2000
523.8'022'3–dc21 00–059909

ISBN 978-0-521-17363-6 Paperback

Cambridge University Press has no responsibility for the persistence or accuracy of URLs for external or third-party internet websites referred to in this publication, and does not guarantee that any content on such websites is, or will remain, accurate or appropriate.

# THE MOON

The Moon is, apart from the Sun, the brightest object in the sky. Although the Sun and the Moon appear almost equal in size, they are quite different. The Sun is the central body of our Solar System, and all planets, including the Earth, orbit around it. The Sun measures 1.4 million kilometers across, and lies at a distance of roughly 150,000,000 kilometers. The Moon is much smaller and measures 'only' 3,476 kilometers across; approximately one quarter of the Earth's diameter. The Moon lies at an average distance of 384,400 kilometers from the Earth. It orbits, not the Sun, but our own planet, in a little more than 27 days.

Although we often refer to the Moon as 'shining' it does not of itself give any light. It only reflects the light it receives from the Sun. This is the reason why the appearance of the Moon changes as it orbits the Earth. This aspect of the Moon, sometimes visible as a thin crescent in the western sky, after sunset, and sometimes as a full disk, lightening up the middle of the night, is confusing to many people. The reason for this can be best explained in a diagram (figure 1).

The illustration is not drawn to scale, but shows you what happens. The Earth is at the center and the Moon's orbit is drawn as a circle. During its orbit around the Earth we see a different portion of the illuminated side of the Moon's surface. (In the figure the red arrows indicate our line of sight from the Earth.)

When the Moon is approximately between the Sun and the Earth, we see only its dark side. We call this a new Moon. The Moon is not visible at all. After one or two days we see a small crescent in the evening sky; a part of the illuminated side is peeping around the edge. After almost a week, half of the disk is lit and we call this the first quarter. Another week later we see the complete disk. This is a full Moon. Next comes the last quarter, and then back to new Moon again. From one new Moon to the next takes about 29.5 days, fully two days longer than it takes the Moon to orbit the Earth. The reason for this is that, in the time the Moon revolves around

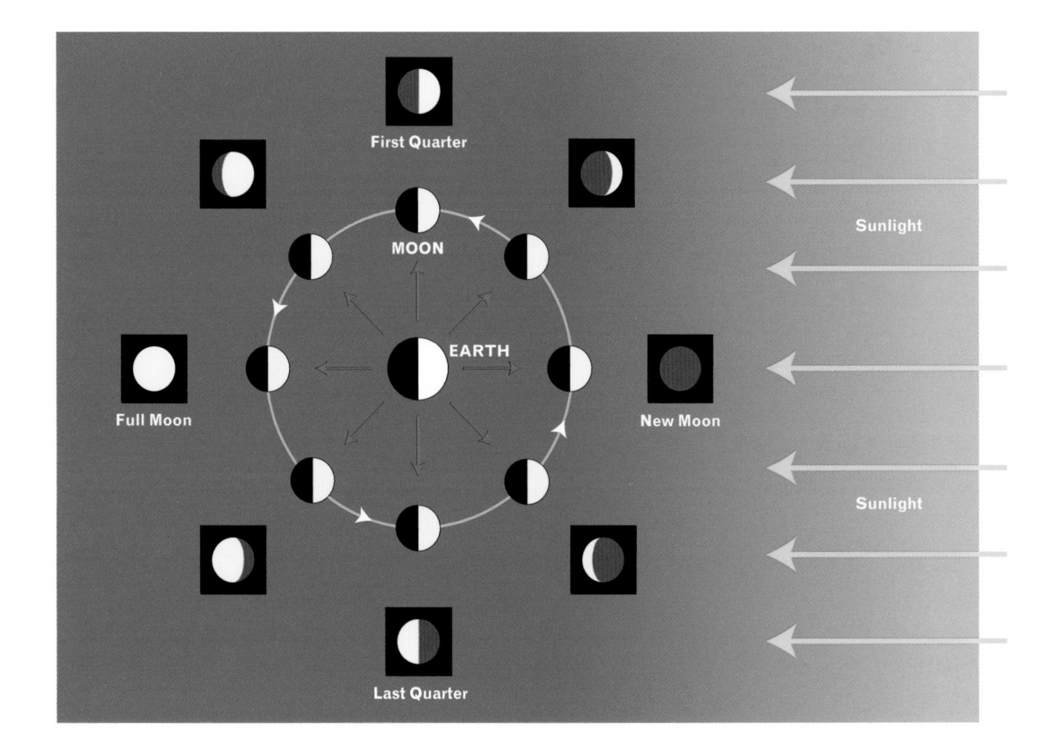


Figure 1

the Earth, the Earth also moves in its orbit around the Sun. In the 27 days it takes the Moon to orbit the Earth, the Sun's position in the sky also changes, moving in the same direction the Moon moves. It takes the Moon more than two days to 'catch up' with the Sun.

## Observing the Moon

Even with a simple pair of binoculars you can see interesting features on the Moon's surface, and a small telescope will reveal even more details of our neighbor in space. The best time to watch the Moon is not when it is full, but rather around its first or last quarter. Then the Moon is illuminated by the Sun from one side and especially near the terminator, the line dividing the lit and the unlit halves of the Moon, there is strong relief, because the surface is illuminated from a very low angle, resulting in long shadows. At full Moon you do not see any relief, since you are then looking from approximately the same direction as the Sun's rays come from. But full Moon is an ideal time to study the differences between the dark and light areas of the surface.

On the Moon maps (on pages 4–5 and 6–7), you can identify the craters and other small features, by referring to the numbers in the list alongside the maps. To help you see which crater a number is referring to, a small black dot is placed in its center. On the first double-page Moon map (map A) the features are listed in numerical order, and for your convenience they are repeated in alphabetical order on the second spread, with the mirror-reversed map (map B). The reason the Moon is shown in two different ways is explained below and on page 3.

Larger features, like mountain ranges and the darker areas called *maria,* are labeled directly on the map. *Maria* is the plural form of the Latin word *mare*, meaning sea. The first observers who believed that these dark areas on the Moon really were seas gave the name. Although we now know there are no seas on the Moon, the name persists, as also do the names *lacus* (lake) and *oceanus* (ocean).

Most craters on the Moon are believed to be the result of the impact of meteors: pieces of rock and metal from space. The Earth is well protected against the impact of meteors by the atmosphere, which causes meteors to burn and vaporize. We call that a 'falling star' or 'shooting star', though it is not a star at all. Only fragments of large meteors reach the surface; we call these fragments meteorites. But the Moon does not have an atmosphere, so every meteor captured by the Moon's gravity will crash into the surface.

Because the Moon rotates 360° on its axis in exactly the same time that it takes to complete one orbit around the Earth, we always see the same side of the Moon. However the Moon's orbit is inclined about 5° to the *ecliptic* (page 37), making it move slightly above and below the plane of the Earth's orbit around the Sun, and the Moon's own spin axis is also tilted about 1.5°. The combined result is that we can occasionally see about 6.5° 'over' the North and South Poles of the Moon. Moreover, since the Moon's orbit is not really a circle, but an ellipse, it does not move at a constant speed, though its rotation speed remains the same. Thus, as seen from the Earth, it moves a little from left to right, as if it were shaking its head very slowly. Therefore, we can look around the edges, by up to 7°. Sometimes Mare Crisium (in the northeastern quadrant of the Moon) appears very close to the edge, and sometimes it is closer to the center, and has a less elliptical appearance. The elliptical appearance of Mare Crisium, as well as those of craters close to the visible edge, is of course caused by perspective.

## Different orientations of the Moon

Naked-eye observers living in the northern hemisphere see the Moon with north up and south down. That is also the way they will see it in a pair of binoculars. However, using an astronomical telescope will usually show the Moon 'upside down'. That is why map A, on pages 4–5, shows the Moon with south at the top, so that will be practical for most observers using a telescope. For naked-eye or binocular observations the map has to be held upside down.

When you are living in the southern hemisphere, it will be the other way around. The map can be used directly for naked-eye and binocular observations, but when you use a telescope, you have to rotate

Milky Way Galaxy. Since our Milky Way Galaxy is a large, flat disk-like formation, we see most stars when we look along the plane of the Milky Way. All the light of the millions of stars that we do not see with the naked eye forms a cloud-like band that is easily visible when you are outside big cities and away from light-polluted areas.

The eight seasonal sky maps are constructed on a stereographic projection. Although this projection has the disadvantage that the scale increases from the center outward, its main advantage is that the shapes of the constellations and star groups are not distorted, making it easier to regognize them. When looking at the sky, there is a strange phenomenon, called the *Moon illusion*, which makes objects appear larger when close to the horizon than when they are high in the sky. This is, however, only a trick of the mind. We believe objects near the horizon to be at a greater distance than objects directly above our heads. The stereographic projection is in a way in harmony with this: constellations near the horizon appear larger!

## Choosing the right map

There are eight 'seasonal' maps, two for each season, placed side-by-side on facing pages. The maps on the left-hand pages are for observers in the northern hemisphere and those on the right-hand pages are for southern observers.

Each map has four different horizons, labeled to show you the latitudes on Earth for which the horizon is exact. Depending on the latitude on Earth at which you are living, you can select your correct horizon. A few degrees more or less will not make too much difference, when you are looking at the sky.

The first map, A, is for January 15, at 10 p.m., the next, B, is for April 15, also at 10 p.m. Map C is for July and D for October, again at 10 p.m. You will notice that the *winter* map for northern observers is coupled with the *summer* map for southern observers. This is because when it is winter in the northern hemisphere of the Earth, it is summer 'down under'.

The maps can be used on other dates and times as well, as is shown in table A. Let's say you want to

| | Local Time | 5 p.m. | 6 p.m. | 7 p.m. | 8 p.m. | 9 p.m. | 10 p.m. | 11 p.m. | Midnight | 1 a.m. | 2 a.m. | 3 a.m. | 4 a.m. | 5 a.m. | 6 a.m. | 7 a.m. |
|---|---|---|---|---|---|---|---|---|---|---|---|---|---|---|---|---|
| | **DST** | 6 | 7 | 8 | 9 | 10 | 11 | M | 1 | 2 | 3 | 4 | 5 | 6 | 7 | 8 |
| January | 1 | **D** | | | | | | **A** | | | | | | **B** | | |
| January | 15 | | | | | | **A** | | | | | | **B** | | | |
| February | 1 | | | | | **A** | | | | | | **B** | | | | |
| February | 15 | | | | **A** | | | | | | **B** | | | | | |
| March | 1 | | | **A** | | | | | | **B** | | | | | | **C** |
| March | 15 | | **A** | | | | | | **B** | | | | | | **C** | |
| April | 1 | **A** | | | | | | **B** | | | | | | **C** | | |
| April | 15 | | | | | | **B** | | | | | | **C** | | | |
| May | 1 | | | | | **B** | | | | | | **C** | | | | |
| May | 15 | | | | **B** | | | | | | **C** | | | | | |
| June | 1 | | | **B** | | | | | | **C** | | | | | | **D** |
| June | 15 | | **B** | | | | | | **C** | | | | | | **D** | |
| July | 1 | **B** | | | | | | **C** | | | | | | **D** | | |
| July | 15 | | | | | | **C** | | | | | | **D** | | | |
| August | 1 | | | | | **C** | | | | | | **D** | | | | |
| August | 15 | | | | **C** | | | | | | **D** | | | | | |
| September | 1 | | | **C** | | | | | | **D** | | | | | | **A** |
| September | 15 | | **C** | | | | | | **D** | | | | | | **A** | |
| October | 1 | **C** | | | | | | **D** | | | | | | **A** | | |
| October | 15 | | | | | | **D** | | | | | | **A** | | | |
| November | 1 | | | | | **D** | | | | | | **A** | | | | |
| November | 15 | | | | **D** | | | | | | **A** | | | | | |
| December | 1 | | | **D** | | | | | | **A** | | | | | | **B** |
| December | 15 | | **D** | | | | | | **A** | | | | | | **B** | |

**Table A** *Seasonal sky maps*

look at the sky in the middle of January, not in the evening, but later in the night. The table tells you that around 4 a.m. you can use map B. Keep in mind that one hour of time difference does not change the sky completely. As already said, during the night the stars move slowly from east to west along the sky, so in practice map B can be used without any problems from let's say 3 a.m. to 5 a.m. For your convenience the local time is also given for DST (Daylight Saving Time).

Another help for finding the right map is the band at the bottom of each chart. Here you see a number of months in the first row, and below that two time lines, one for regular time and one for DST. Above that you also see a key to the magnitudes. The word *magnitude* refers to the brightness of a star and is further explained in the introduction to the star charts, on page 31.

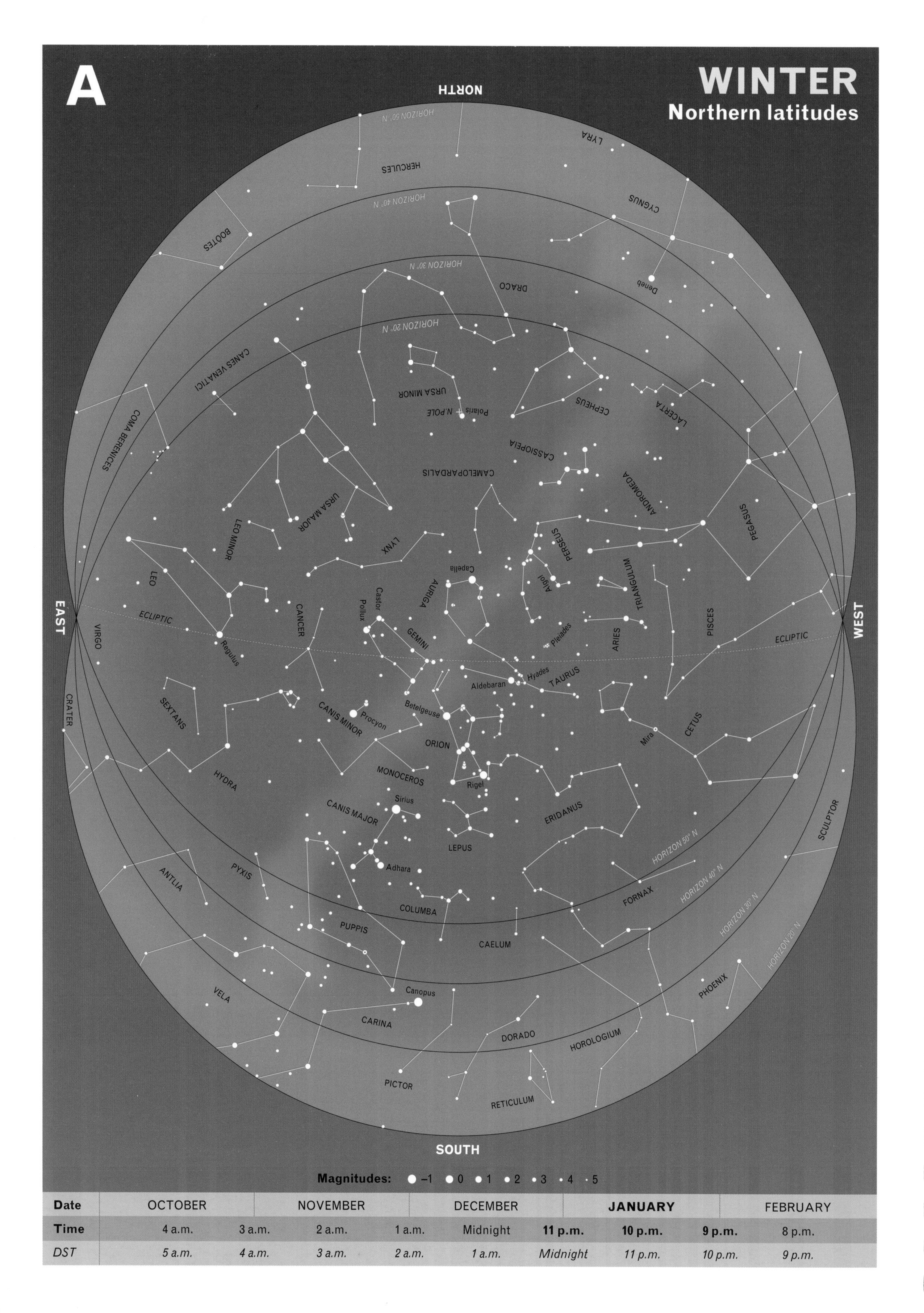


| Date | OCTOBER | | NOVEMBER | | DECEMBER | | JANUARY | | FEBRUARY |
|---|---|---|---|---|---|---|---|---|---|
| Time | 4 a.m. | 3 a.m. | 2 a.m. | 1 a.m. | Midnight | **11 p.m.** | **10 p.m.** | **9 p.m.** | 8 p.m. |
| *DST* | *5 a.m.* | *4 a.m.* | *3 a.m.* | *2 a.m.* | *1 a.m.* | *Midnight* | *11 p.m.* | *10 p.m.* | *9 p.m.* |

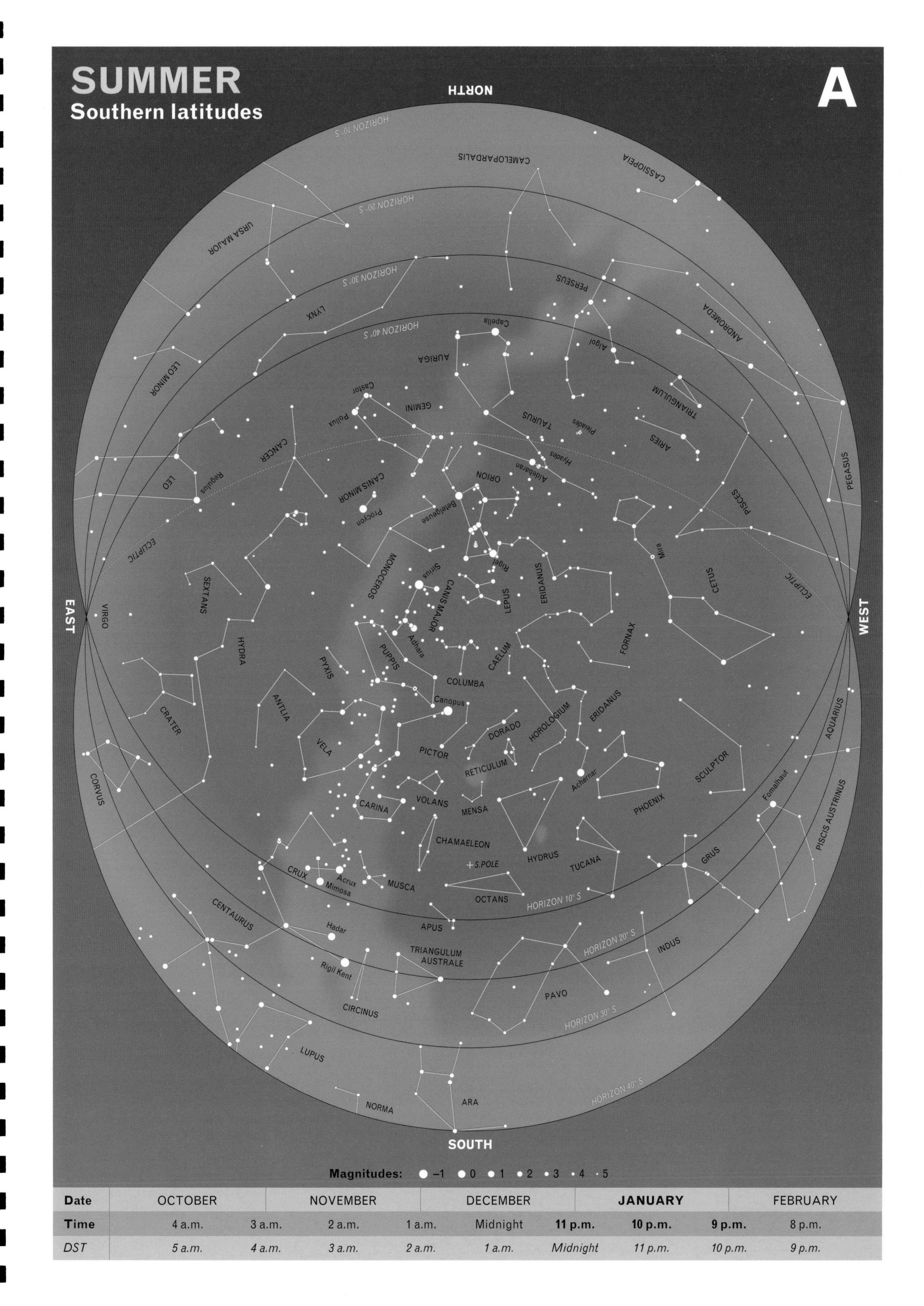


| Date | OCTOBER | | NOVEMBER | | DECEMBER | | **JANUARY** | | FEBRUARY |
|---|---|---|---|---|---|---|---|---|---|
| **Time** | 4 a.m. | 3 a.m. | 2 a.m. | 1 a.m. | Midnight | **11 p.m.** | **10 p.m.** | **9 p.m.** | 8 p.m. |
| *DST* | *5 a.m.* | *4 a.m.* | *3 a.m.* | *2 a.m.* | *1 a.m.* | *Midnight* | *11 p.m.* | *10 p.m.* | *9 p.m.* |

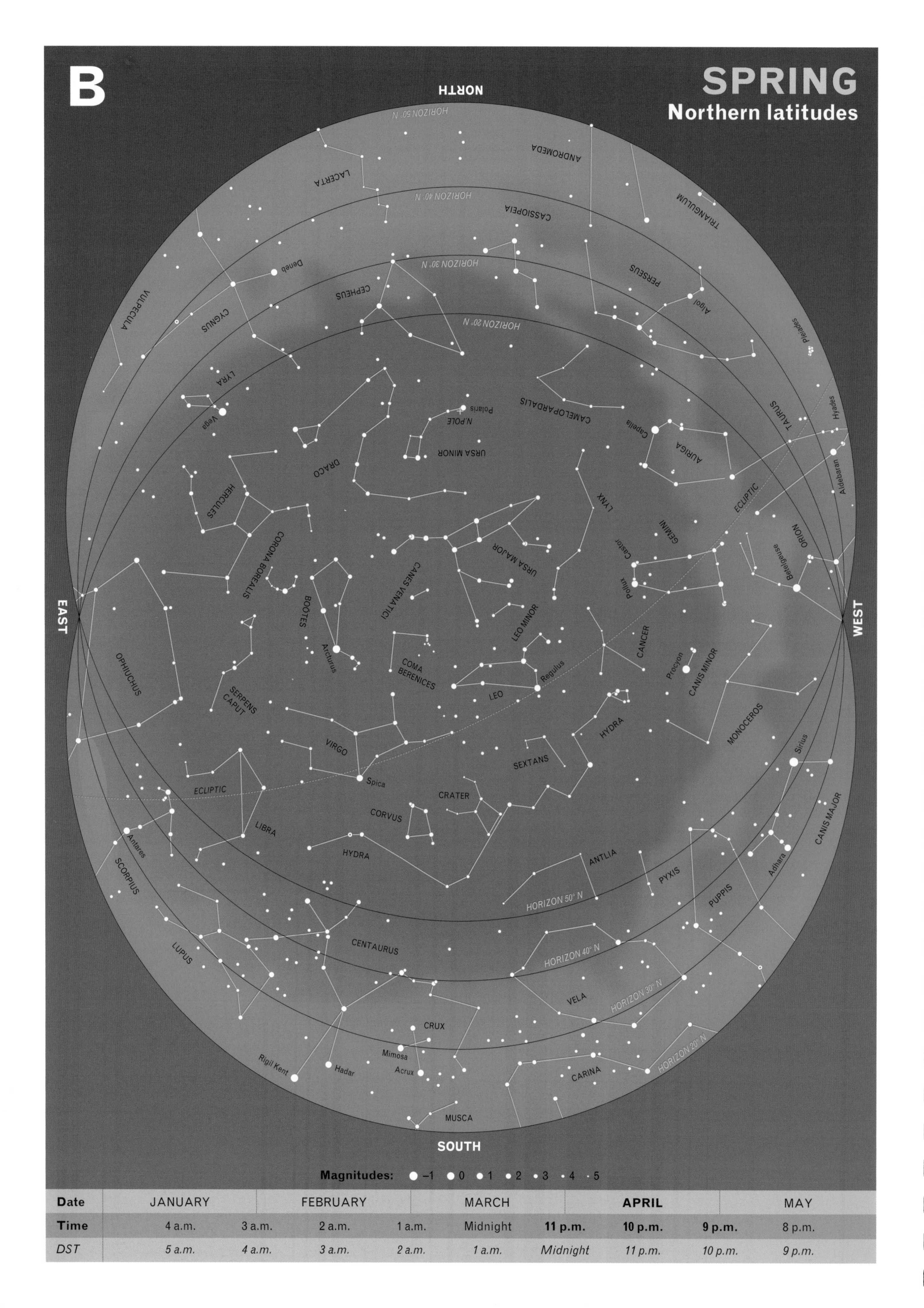


| Date | JANUARY | | FEBRUARY | | MARCH | | APRIL | | MAY |
|---|---|---|---|---|---|---|---|---|---|
| Time | 4 a.m. | 3 a.m. | 2 a.m. | 1 a.m. | Midnight | **11 p.m.** | **10 p.m.** | **9 p.m.** | 8 p.m. |
| *DST* | *5 a.m.* | *4 a.m.* | *3 a.m.* | *2 a.m.* | *1 a.m.* | *Midnight* | *11 p.m.* | *10 p.m.* | *9 p.m.* |

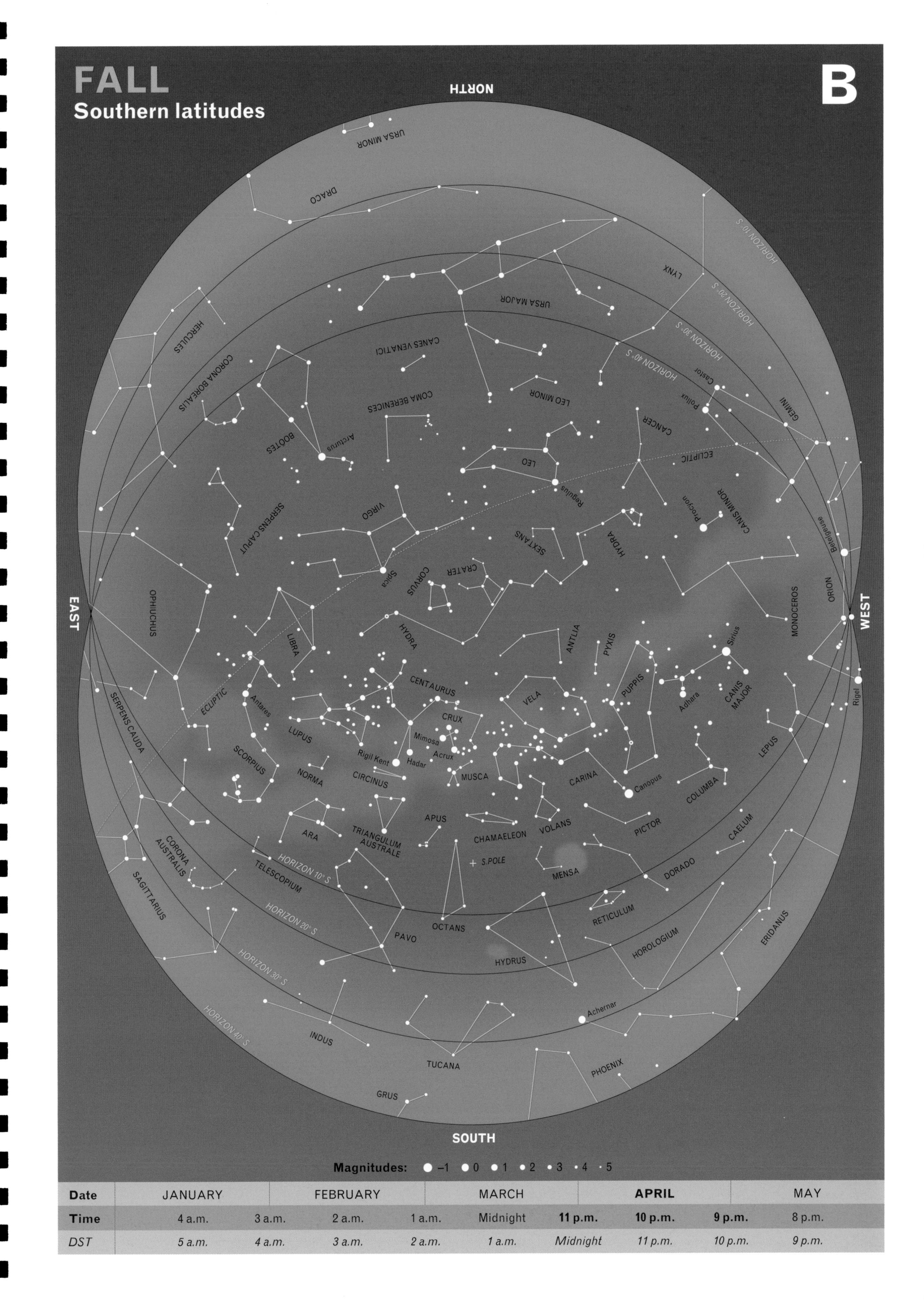


| Date | JANUARY | | FEBRUARY | | MARCH | | **APRIL** | | MAY |
|---|---|---|---|---|---|---|---|---|---|
| **Time** | 4 a.m. | 3 a.m. | 2 a.m. | 1 a.m. | Midnight | **11 p.m.** | **10 p.m.** | **9 p.m.** | 8 p.m. |
| *DST* | *5 a.m.* | *4 a.m.* | *3 a.m.* | *2 a.m.* | *1 a.m.* | *Midnight* | *11 p.m.* | *10 p.m.* | *9 p.m.* |

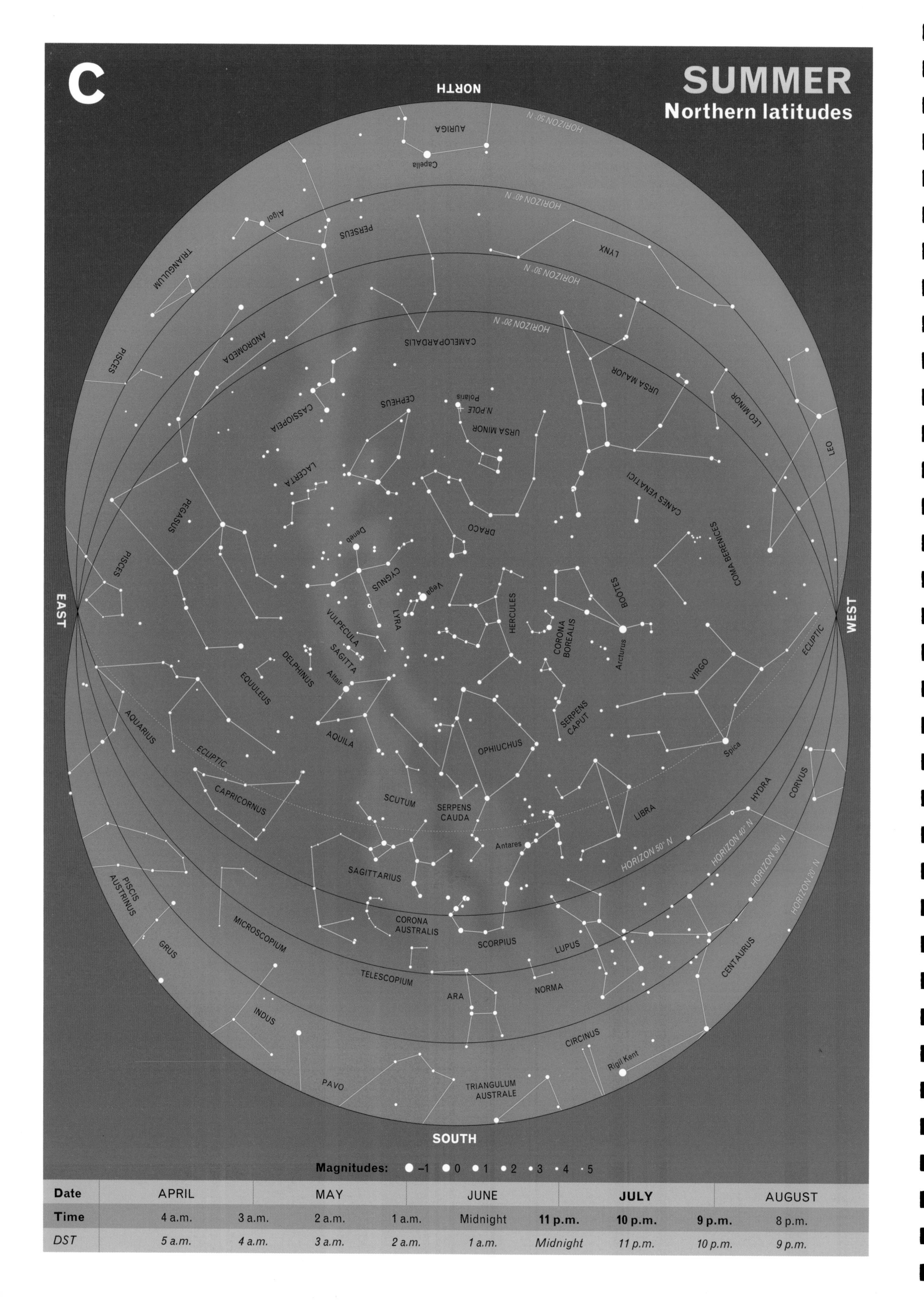


| Date | APRIL | | MAY | | JUNE | | JULY | | AUGUST |
|---|---|---|---|---|---|---|---|---|---|
| **Time** | 4 a.m. | 3 a.m. | 2 a.m. | 1 a.m. | Midnight | **11 p.m.** | **10 p.m.** | **9 p.m.** | 8 p.m. |
| *DST* | *5 a.m.* | *4 a.m.* | *3 a.m.* | *2 a.m.* | *1 a.m.* | *Midnight* | *11 p.m.* | *10 p.m.* | *9 p.m.* |

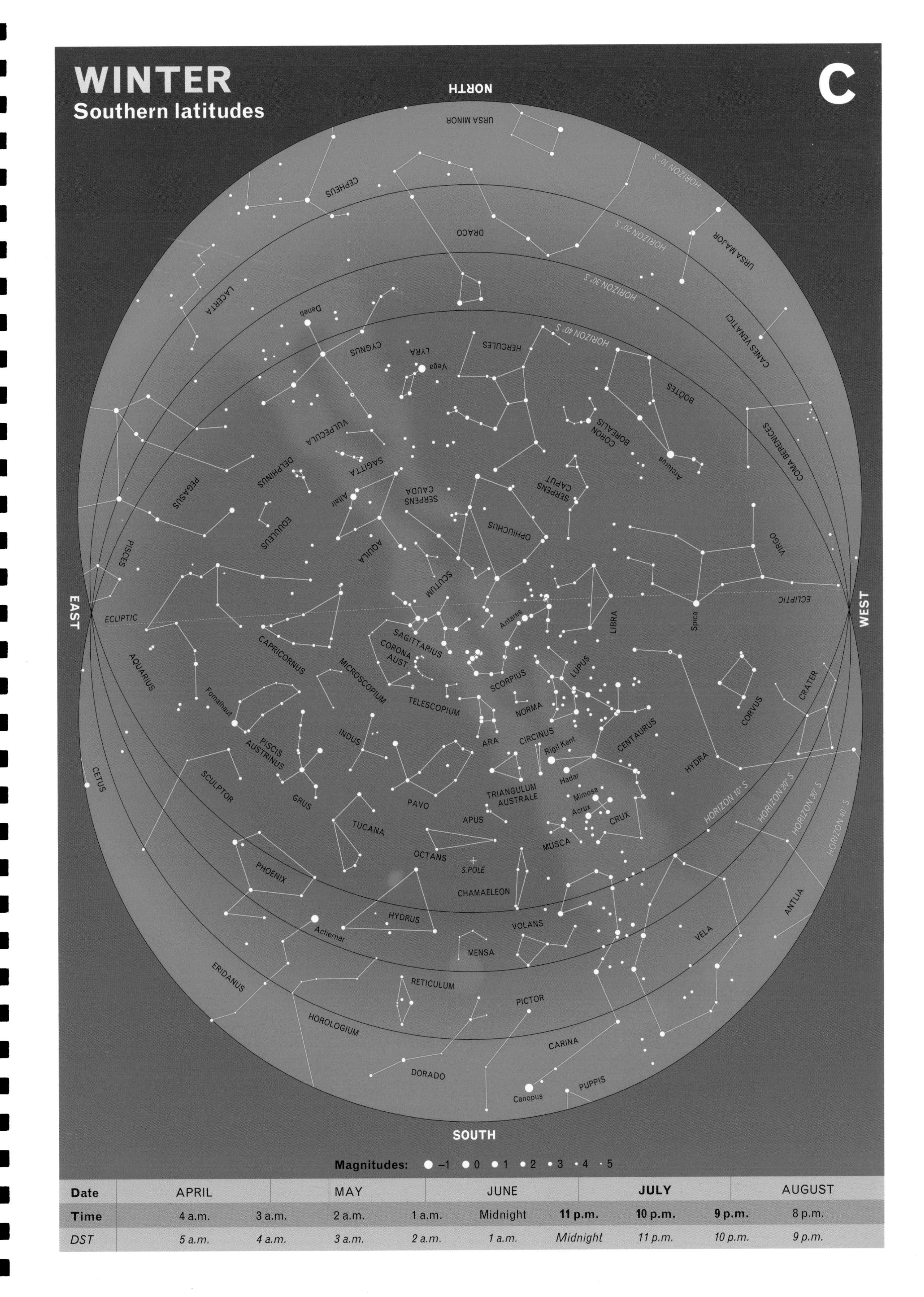


| Date | APRIL | | MAY | | JUNE | | JULY | | AUGUST |
|---|---|---|---|---|---|---|---|---|---|
| Time | 4 a.m. | 3 a.m. | 2 a.m. | 1 a.m. | Midnight | **11 p.m.** | **10 p.m.** | **9 p.m.** | 8 p.m. |
| *DST* | *5 a.m.* | *4 a.m.* | *3 a.m.* | *2 a.m.* | *1 a.m.* | *Midnight* | *11 p.m.* | *10 p.m.* | *9 p.m.* |

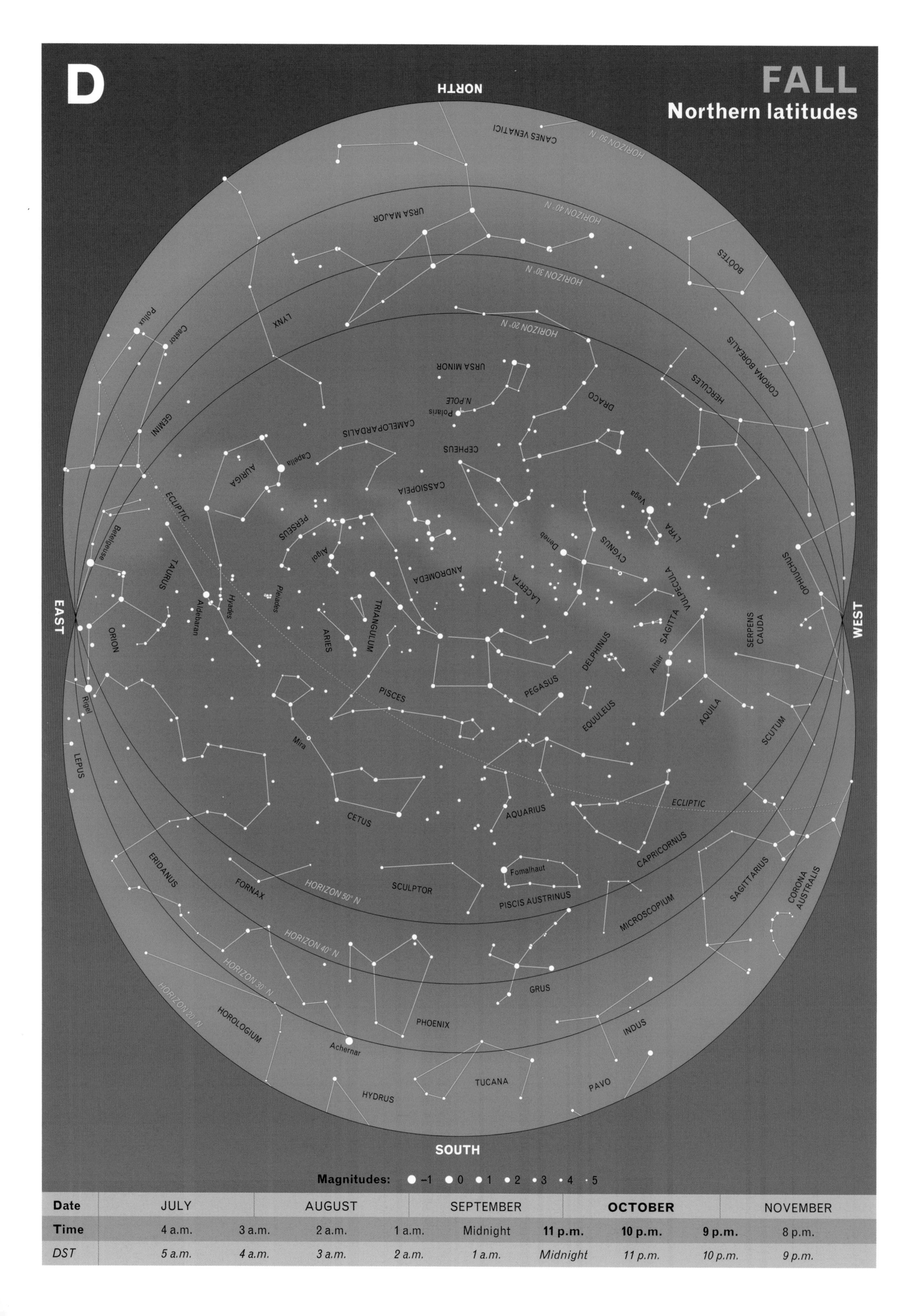


| Date | JULY | | AUGUST | | SEPTEMBER | | OCTOBER | | NOVEMBER |
|---|---|---|---|---|---|---|---|---|---|
| Time | 4 a.m. | 3 a.m. | 2 a.m. | 1 a.m. | Midnight | **11 p.m.** | **10 p.m.** | **9 p.m.** | 8 p.m. |
| *DST* | *5 a.m.* | *4 a.m.* | *3 a.m.* | *2 a.m.* | *1 a.m.* | *Midnight* | *11 p.m.* | *10 p.m.* | *9 p.m.* |

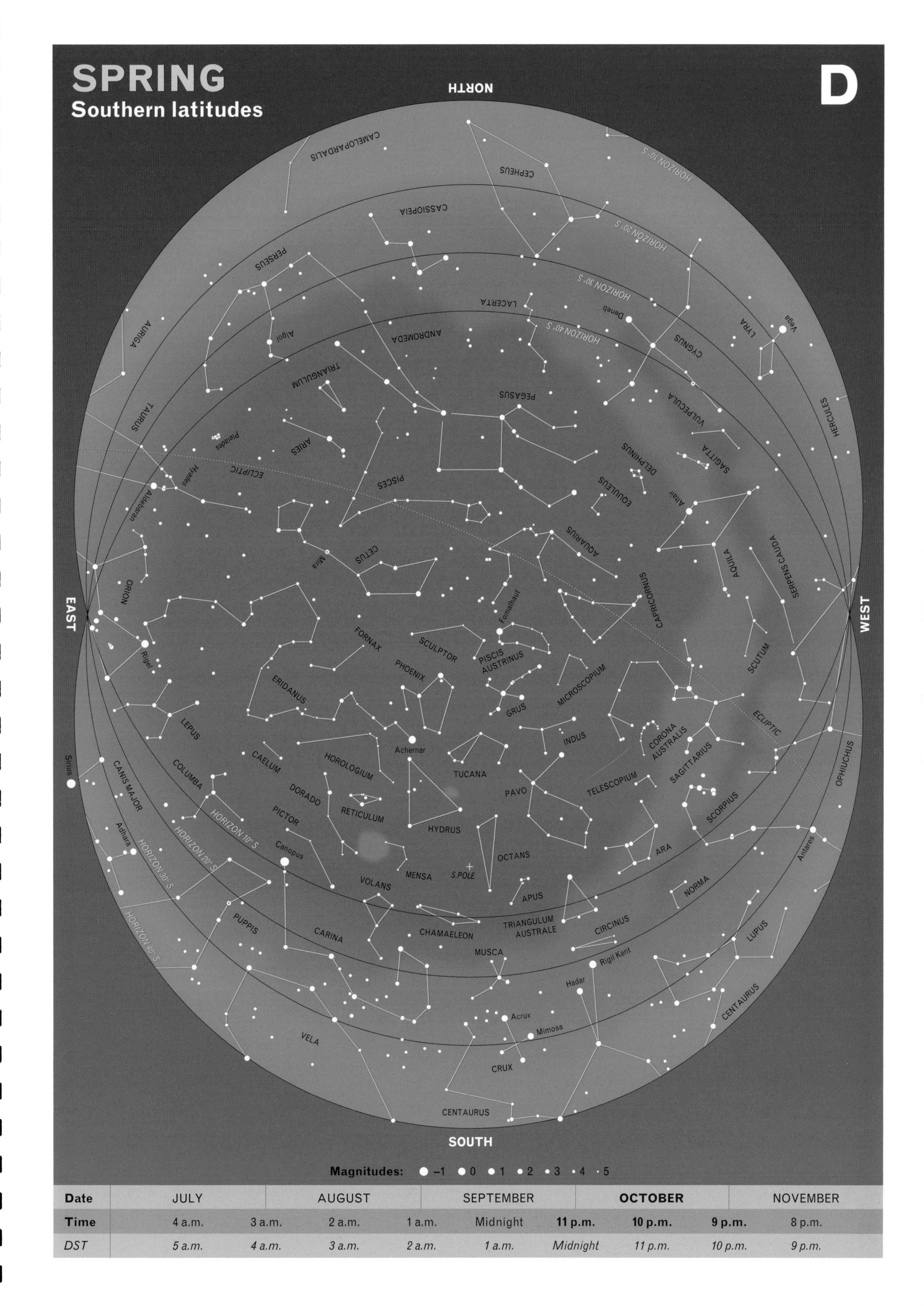


| Date | JULY | | AUGUST | | SEPTEMBER | | OCTOBER | | NOVEMBER |
|---|---|---|---|---|---|---|---|---|---|
| Time | 4 a.m. | 3 a.m. | 2 a.m. | 1 a.m. | Midnight | **11 p.m.** | **10 p.m.** | **9 p.m.** | 8 p.m. |
| *DST* | *5 a.m.* | *4 a.m.* | *3 a.m.* | *2 a.m.* | *1 a.m.* | *Midnight* | *11 p.m.* | *10 p.m.* | *9 p.m.* |

# THE MESSIER OBJECTS

The French astronomer *Charles Messier* (1730–1817) created a list of 103 *deep-sky objects* (star clusters, nebulae and galaxies) that was first published in 1774. Later, other astronomers added a few more objects and today the list contains 110 objects. In the charts on the following six pages and in the star charts on pages 40–79 these objects are labeled with the number from his list with the prefix M (like M42, M78, etc).

Messier was a 'comet hunter', and while scanning the sky from his observatory in Paris he frequently ran into faint objects that could be mistaken for comets. He compiled the list to help him distinguish between the real comets and these permanent objects. Although most of the objects in Messier's list were later included in larger catalogs of deep-sky objects, Messier's list remains immensely popular amongst amateur observers. That is why the charts on the following eight pages focus on these objects.

All Messier objects are clearly plotted on the charts, against a background showing all stars brighter than magnitude 5.5. The meaning of the word *magnitude* is explained on page 31. For the abbreviations used in the column head, you may refer to page 38. The symbols used for the different types of objects are the same as used in the main star charts.

The charts in this chapter also serve as an *index* to the star charts. The area covered by each chart (not including the overlap) is marked with lines in a lighter blue, and the number is clearly shown in a slightly darker color.

Table B provides you with information about all the Messier objects, and they can easily be identified on the four double-page charts (a, b, c and d) on pages 22–29. For your convenience, the chart numbers are listed in the final columns, together with the numbers of the main star charts, where they can be found (1, 2, 3, etc).

**Table B** *The Messier objects*

| Messier number | NGC/IC | Con | RA | | Dec | | Mag | Size | Type * | Common name / note | Chart numbers | |
|---|---|---|---|---|---|---|---|---|---|---|---|---|
| | | | h | m | ° | ' | | ' | | | Messier | Main |
| 1 | 1952 | Tau | 05 | 34.5 | +22 | 01 | 8.4 | 6 × 4 | N | *Crab Nebula* | b | 3 9 |
| 2 | 7089 | Aqr | 21 | 33.5 | −00 | 49 | 6.5 | 12.9 | GC | | c | 13 |
| 3 | 5272 | CVn | 13 | 42.2 | +28 | 23 | 6.4 | 16.2 | GC | | a c | 5 |
| 4 | 6121 | Sco | 16 | 23.6 | −26 | 32 | 5.9 | 26.3 | GC | | c d | 11 12 17 18 |
| 5 | 5904 | Ser | 15 | 18.6 | +02 | 05 | 5.8 | 17.4 | GC | | c | 11 12 |
| 6 | 6405 | Sco | 17 | 40.1 | −32 | 13 | 4.2 | 15 | OC | *Butterfly Cluster* | c d | 18 |
| 7 | 6475 | Sco | 17 | 53.9 | −34 | 49 | 3.3 | 80 | OC | *Ptolemy's Cluster* | c d | 18 |
| 8 | 6523 | Sgr | 18 | 03.8 | −24 | 23 | 5.8 | 90 × 40 | N | *Lagoon Nebula* | c | 12 18 |
| 9 | 6333 | Oph | 17 | 19.2 | −18 | 31 | 7.9 | 9.3 | GC | | c | 12 18 |
| 10 | 6254 | Oph | 16 | 57.1 | −04 | 06 | 6.6 | 15.1 | GC | | c | 12 |
| 11 | 6705 | Sct | 18 | 51.1 | −06 | 16 | 5.8 | 14 | OC | *Wild Duck Cluster* | c | 12 |
| 12 | 6218 | Oph | 16 | 47.2 | −01 | 57 | 6.6 | 14.5 | GC | | c | 12 |
| 13 | 6205 | Her | 16 | 41.7 | +36 | 28 | 5.9 | 16.6 | GC | *Hercules Cluster* | a c | 5 6 |
| 14 | 6402 | Oph | 17 | 37.6 | −03 | 15 | 7.6 | 11.7 | GC | | c | 12 |
| 15 | 7078 | Peg | 21 | 30.0 | +12 | 10 | 6.4 | 12.3 | GC | | c | 13 |
| 16 | 6611 | Ser | 18 | 18.8 | −13 | 47 | 6.0 | 35 × 28 | N+OC | *Eagle Nebula* | c | 12 |
| 17 | 6618 | Sgr | 18 | 20.8 | −16 | 11 | 7 | 46 × 37 | N+OC | *Omega Nebula* | c | 12 |
| 18 | 6613 | Sgr | 18 | 19.9 | −17 | 08 | 6.9 | 9 | OC | | c | 12 18 |
| 19 | 6273 | Oph | 17 | 02.6 | −26 | 16 | 7.2 | 13.5 | GC | | c d | 12 18 |
| 20 | 6514 | Sgr | 18 | 02.6 | −23 | 02 | 8.5 | 29 × 27 | N | *Trifid Nebula* | c | 12 18 |
| 21 | 6531 | Sgr | 18 | 04.6 | −22 | 30 | 5.9 | 13 | OC | | c | 12 18 |
| 22 | 6656 | Sgr | 18 | 36.4 | −23 | 54 | 5.1 | 24 | GC | | c | 12 18 |
| 23 | 6495 | Sgr | 17 | 56.8 | −19 | 01 | 5.5 | 27 | OC | | c | 12 18 |
| 24 | — | Sgr | 18 | 16.9 | −18 | 29 | 4.5 | 90 | — | Star cloud – not a real cluster | c | 12 18 |
| 25 | 4525 (IC) | Sgr | 18 | 31.6 | −19 | 15 | 4.6 | 32 | OC | | c | 12 18 |
| 26 | 6694 | Sct | 18 | 45.2 | −09 | 24 | 8.0 | 15 | OC | | c | 12 |
| 27 | 6853 | Vul | 19 | 59.6 | +22 | 43 | 7.6 | 8 × 4 | PN | *Dumbbell Nebula* | a c | 6 7 12 13 |
| 28 | 6626 | Sgr | 18 | 24.5 | −24 | 52 | 6.9 | 11.2 | GC | | c | 12 18 |
| 29 | 6913 | Cyg | 20 | 23.9 | +38 | 32 | 6.6 | 7 | OC | | a c | 6 7 |
| 30 | 7099 | Cap | 21 | 40.4 | −23 | 11 | 7.5 | 11 | GC | | c d | 13 19 |
| 31 | 224 | And | 00 | 42.7 | +41 | 16 | 3.5 | 178 × 63 | GX | *Andromeda Galaxy* | a b c | 2 7 |
| 32 | 221 | And | 00 | 42.7 | +40 | 52 | 8.2 | 7.6 × 5.8 | GX | Companion to M31 | a b c | 2 7 |
| 33 | 598 | Tri | 01 | 33.9 | +30 | 39 | 5.7 | 62 × 39 | GX | *Triangulum Galaxy* | a b | 2 |
| 34 | 1039 | Per | 02 | 42.0 | +42 | 47 | 5.2 | 35 | OC | | a b | 2 3 |

**Table B** *The Messier objects (continued)*

| Messier number | NGC/IC | Con | RA h | RA m | Dec ° | Dec ' | Mag | Size ' | Type * | Common name / note | Chart numbers Messier | Chart numbers Main |
|---|---|---|---|---|---|---|---|---|---|---|---|---|
| 35 | 2168 | Gem | 06 | 08.9 | +24 | 20 | 5.1 | 28 | OC | | b | 3 9 |
| 36 | 1960 | Aur | 05 | 36.1 | +34 | 08 | 6.0 | 12 | OC | | a b | 3 |
| 37 | 2099 | Aur | 05 | 52.4 | +32 | 33 | 5.6 | 24 | OC | | a b | 3 |
| 38 | 1912 | Aur | 05 | 28.7 | +35 | 50 | 6.4 | 21 | OC | | a b | 3 |
| 39 | 7092 | Cyg | 21 | 32.2 | +48 | 26 | 4.6 | 32 | OC | | a c | 1 6 7 |
| 40 | — | UMa | 12 | 22.4 | +58 | 05 | 8 | | — | Double star | a b c | 1 4 5 |
| 41 | 2287 | CMa | 06 | 47.0 | −20 | 44 | 4.5 | 38 | OC | | b | 9 15 |
| 42 | 1976 | Ori | 05 | 35.4 | −05 | 27 | 4 | 66 × 60 | N | *Great Orion Nebula* | b | 9 |
| 43 | 1982 | Ori | 05 | 35.6 | −05 | 16 | 9 | 20 × 15 | N | Extension of M42 | b | 9 |
| 44 | 2632 | Cnc | 08 | 40.1 | +19 | 59 | 3.1 | 95 | OC | *Praesepe; Beehive Cluster* | a b | 3 4 9 10 |
| 45 | — | Tau | 03 | 47.0 | +24 | 07 | 1.2 | 110 | OC | *Pleiades; Seven Sisters* | a b | 2 3 8 9 |
| 46 | 2437 | Pup | 07 | 41.8 | −14 | 49 | 6.1 | 27 | OC | | b | 9 10 15 16 |
| 47 | 2422 | Pup | 07 | 36.6 | −14 | 30 | 4.4 | 30 | OC | | b | 9 10 15 16 |
| 48 | 2548 | Hya | 08 | 13.8 | −05 | 48 | 5.8 | 54 | OC | | b | 9 10 |
| 49 | 4472 | Vir | 12 | 29.8 | +08 | 00 | 8.4 | 8.9 × 7.4 | GX | | b c | 10 11 |
| 50 | 2323 | Mon | 07 | 03.2 | −08 | 20 | 5.9 | 16 | OC | | b | 9 |
| 51 | 5194, 5195 | CVn | 13 | 29.9 | +47 | 12 | 8.4 | 11.0 × 7.8 | GX | *Whirlpool Galaxy* | a c | 4 5 |
| 52 | 7654 | Cas | 23 | 24.2 | +61 | 35 | 6.9 | 13 | OC | | a | 1 2 7 |
| 53 | 5024 | Com | 13 | 12.9 | +18 | 10 | 7.7 | 12.6 | GC | | c | 5 11 |
| 54 | 6715 | Sgr | 18 | 55.1 | −30 | 29 | 7.7 | 9.1 | GC | | c d | 18 19 |
| 55 | 6809 | Sgr | 19 | 40.0 | −38 | 58 | 7.0 | 19.0 | GC | | c d | 18 19 |
| 56 | 6779 | Lyr | 19 | 16.6 | +30 | 11 | 8.2 | 7.1 | GC | | a c | 6 7 |
| 57 | 6720 | Lyr | 18 | 53.6 | +33 | 02 | 9.7 | 1.2 | PL | *Ring Nebula* | a c | 6 7 |
| 58 | 4579 | Vir | 12 | 37.7 | +11 | 49 | 9.8 | 5.4 × 4.4 | GX | | b c | 10 11 |
| 59 | 4621 | Vir | 12 | 42.0 | +11 | 39 | 9.8 | 5.1 × 3.4 | GX | | b c | 11 |
| 60 | 4649 | Vir | 12 | 43.7 | +11 | 33 | 8.8 | 7.2 × 6.2 | GX | | b c | 11 |
| 61 | 4303 | Vir | 12 | 21.9 | +04 | 28 | 9.7 | 6.0 × 5.5 | GX | | b c | 10 11 |
| 62 | 6266 | Oph | 17 | 01.2 | −30 | 07 | 6.6 | 14.1 | GC | | c d | 17 18 |
| 63 | 5055 | CVn | 13 | 15.8 | +41 | 02 | 8.6 | 12.3 × 7.6 | GX | *Sunflower Galaxy* | a c | 4 5 |
| 64 | 4826 | Com | 12 | 56.7 | +21 | 41 | 8.5 | 9.3 × 5.4 | GX | *Black-Eye Galaxy* | b c | 5 11 |
| 65 | 3623 | Leo | 11 | 18.9 | +13 | 05 | 9.3 | 10.0 × 3.3 | GX | | b c | 10 11 |
| 66 | 3627 | Leo | 11 | 20.2 | +12 | 59 | 9.0 | 8.7 × 4.4 | GX | | b c | 10 11 |
| 67 | 2682 | Cnc | 08 | 50.4 | +11 | 49 | 6.9 | 30 | OC | | a b | 10 |
| 68 | 4590 | Hya | 12 | 39.5 | −26 | 45 | 8.2 | 12.0 | GC | | b c d | 10 11 16 17 |
| 69 | 6637 | Sgr | 18 | 31.4 | −32 | 21 | 7.7 | 7.1 | GC | | c d | 18 |
| 70 | 6681 | Sgr | 18 | 43.2 | −32 | 18 | 8.1 | 7.8 | GC | | c d | 18 |
| 71 | 6838 | Sge | 19 | 53.8 | +18 | 47 | 8.3 | 7.2 | GC | | c | 6 7 12 13 |
| 72 | 6981 | Aqr | 20 | 53.5 | −12 | 32 | 9.4 | 5.9 | GC | | c d | 13 |
| 73 | 6997 | Aqr | 20 | 58.9 | −12 | 38 | 8.9 | 2.8 | — | Not a real cluster | c d | 13 |
| 74 | 628 | Psc | 01 | 36.7 | +15 | 47 | 9.2 | 10.2 × 9.5 | GX | | c | 8 |
| 75 | 6864 | Sgr | 20 | 06.1 | −21 | 55 | 8.6 | 6.0 | GC | | c d | 12 13 18 19 |
| 76 | 650, 651 | Per | 01 | 42.4 | +51 | 34 | 12.2 | 2 × 1 | PL | *Little Dumbbell Nebula* | a b | 1 2 3 7 |
| 77 | 1068 | Cet | 02 | 42.7 | −00 | 01 | 8.8 | 6.9 × 5.9 | GX | | b | 8 |
| 78 | 2068 | Ori | 05 | 46.7 | +00 | 03 | 8 | 8 × 6 | N | | b | 9 |
| 79 | 1904 | Lep | 05 | 24.5 | −24 | 33 | 8.0 | 8.7 | GC | | b | 9 15 |
| 80 | 6093 | Sco | 16 | 17.0 | −22 | 59 | 7.2 | 8.9 | GC | | c d | 11 12 17 18 |
| 81 | 3031 | UMa | 09 | 55.6 | +69 | 04 | 6.9 | 25.7 × 14.1 | GX | *Bode's Galaxy* | a | 1 4 |
| 82 | 3034 | UMa | 09 | 55.8 | +69 | 41 | 8.4 | 11.2 × 4.6 | GX | *Cigar Galaxy* | a | 1 4 |
| 83 | 5236 | Hya | 13 | 37.0 | −29 | 52 | 8.2 | 11.2 × 10.2 | GX | *Southern Pinwheel Galaxy* | c d | 17 |
| 84 | 4374 | Vir | 12 | 25.1 | +12 | 53 | 9.3 | 5.0 × 4.4 | GX | | b c | 4 10 11 |
| 85 | 4382 | Com | 12 | 25.4 | +18 | 11 | 9.2 | 7.1 × 5.2 | GX | | b c | 4 5 10 11 |
| 86 | 4406 | Vir | 12 | 26.2 | +12 | 57 | 9.2 | 7.5 × 5.5 | GX | | b c | 4 10 11 |
| 87 | 4486 | Vir | 12 | 30.8 | +12 | 24 | 8.6 | 7.2 × 6.8 | GX | Virgo A | b c | 10 11 |
| 88 | 4501 | Com | 12 | 32.0 | +14 | 25 | 9.5 | 6.9 × 3.9 | GX | | b c | 5 10 11 |
| 89 | 4552 | Vir | 12 | 35.7 | +12 | 33 | 9.8 | 4.2 × 4.2 | GX | | b c | 10 11 |
| 90 | 4569 | Vir | 12 | 36.8 | +13 | 10 | 9.5 | 9.5 × 4.7 | GX | | b c | 10 11 |
| 91 | 4548 | Com | 12 | 35.4 | +14 | 30 | 10.2 | 5.4 × 4.4 | GX | | b c | 5 10 11 |
| 92 | 6341 | Her | 17 | 17.1 | +43 | 08 | 6.5 | 11.2 | GC | | a c | 5 6 |
| 93 | 2447 | Pup | 07 | 44.6 | −23 | 52 | 6.2 | 22 | OC | | b d | 9 10 15 16 |
| 94 | 4736 | CVn | 12 | 50.9 | +41 | 07 | 8.2 | 11.0 × 9.1 | GX | | a b c | 4 5 |
| 95 | 3351 | Leo | 10 | 44.0 | +11 | 42 | 9.7 | 7.4 × 5.1 | GX | | b | 10 |
| 96 | 3368 | Leo | 10 | 46.8 | +11 | 49 | 9.2 | 7.1 × 5.1 | GX | | b | 10 |
| 97 | 3587 | UMa | 11 | 14.8 | +55 | 01 | 12.0 | 3.2 | PL | *Owl Nebula* | a b c | 1 4 5 |
| 98 | 4192 | Com | 12 | 13.8 | +14 | 54 | 10.1 | 9.5 × 3.2 | GX | | b c | 4 5 10 11 |
| 99 | 4254 | Com | 12 | 18.8 | +14 | 25 | 9.8 | 5.4 × 4.8 | GX | | b c | 4 5 10 11 |
| 100 | 4321 | Com | 12 | 22.9 | +15 | 49 | 9.4 | 6.9 × 6.2 | GX | | b c | 4 5 10 11 |
| 101 | 5457 | UMa | 14 | 03.2 | +54 | 21 | 7.7 | 26.9 × 26.3 | GX | *Pinwheel Galaxy* | a c | 1 4 5 6 |
| 102 | 5866 | Dra | 15 | 06.5 | +55 | 46 | 10.0 | 5.2 × 2.3 | GX | *Spindle Galaxy* | a c | 1 5 6 |
| 103 | 581 | Cas | 01 | 33.2 | +60 | 42 | 7.4 | 6 | OC | | a b | 1 2 3 7 |
| 104 | 4594 | Vir | 12 | 40.0 | −11 | 37 | 8.3 | 8.9 × 4.1 | GX | *Sombrero Galaxy* | b c | 10 11 |
| 105 | 3379 | Leo | 10 | 47.8 | +12 | 35 | 9.3 | 4.5 × 4.0 | GX | | b | 10 |
| 106 | 4258 | CVn | 12 | 19.0 | +47 | 18 | 8.3 | 18.2 × 7.9 | GX | | a b c | 4 5 |
| 107 | 6171 | Oph | 16 | 32.5 | −13 | 03 | 8.1 | 10.0 | GC | | c | 11 12 |
| 108 | 3556 | UMa | 11 | 11.5 | +55 | 40 | 10.1 | 8.3 × 2.5 | GX | | a b c | 1 4 5 |
| 109 | 3992 | UMa | 11 | 57.6 | +53 | 23 | 9.8 | 7.6 × 4.9 | GX | | a b c | 1 4 5 |
| 110 | 205 | And | 00 | 40.4 | +41 | 41 | 8.0 | 17.4 × 9.8 | GX | Companion to M31 | a b c | 2 7 |

* OC = Open star cluster   GC = Globular cluster   N = (Diffuse) nebula   PL = Planetary nebula   GX = Galaxy

*(You can find a description of these different types of objects on pages 36 and 37.)*

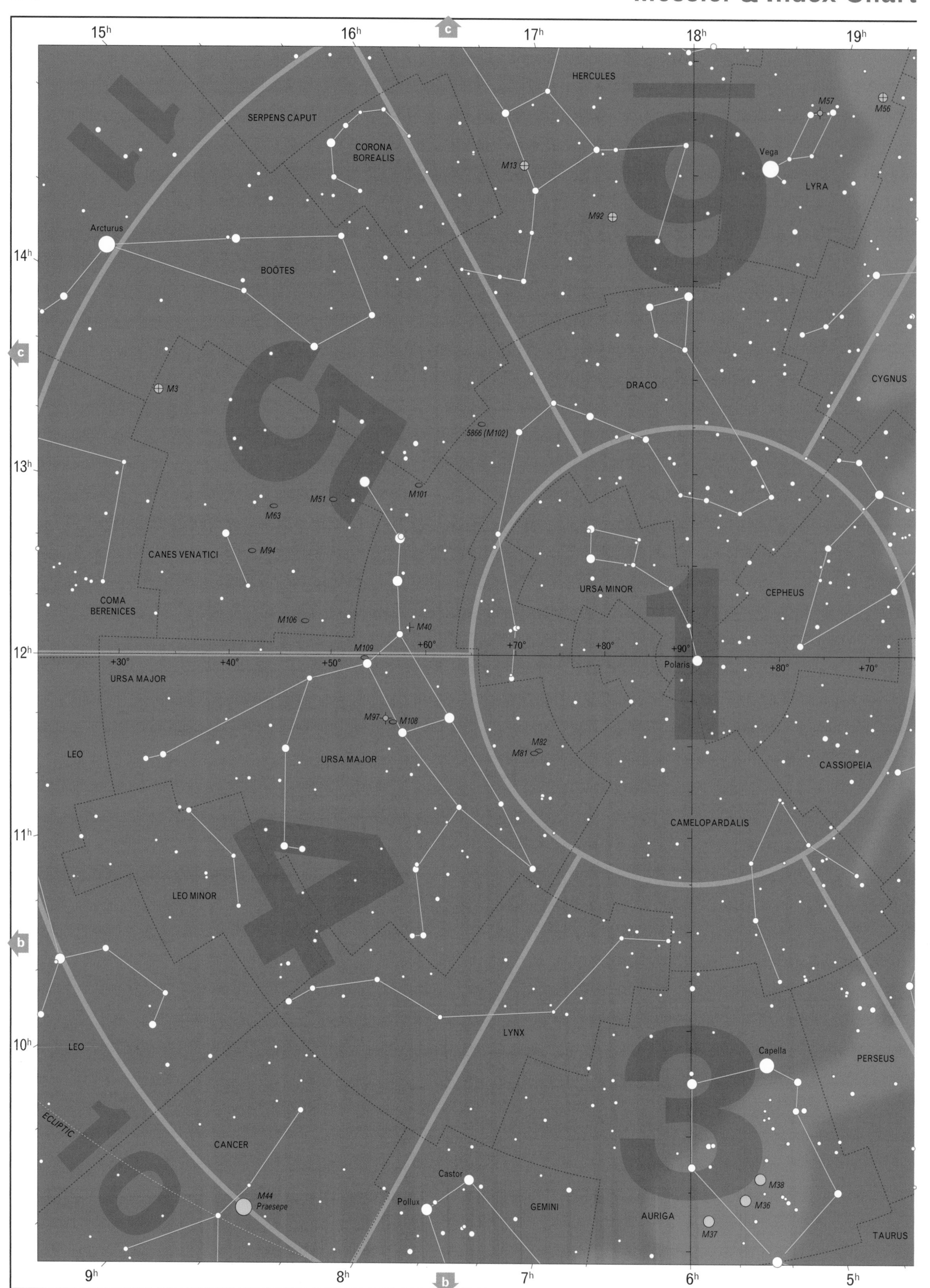
15h
16h
17h
18h
19h
14h
13h
12h
11h
10h
9h
8h
7h
6h
5h
c
b
HERCULES
SERPENS CAPUT
CORONA BOREALIS
M57
M56
Vega
LYRA
M13
M92
Arcturus
BOÖTES
DRACO
CYGNUS
M3
5866 (M102)
M101
M51
M63
M94
CANES VENATICI
COMA BERENICES
URSA MINOR
CEPHEUS
M106
M40
M109
+30°
+40°
+50°
+60°
+70°
+80°
+90°
Polaris
URSA MAJOR
M97
M108
M82
M81
LEO
CASSIOPEIA
CAMELOPARDALIS
LEO MINOR
LYNX
Capella
PERSEUS
ECLIPTIC
CANCER
Castor
Pollux
M44
Praesepe
GEMINI
AURIGA
M38
M36
M37
TAURUS

## Northern Sky

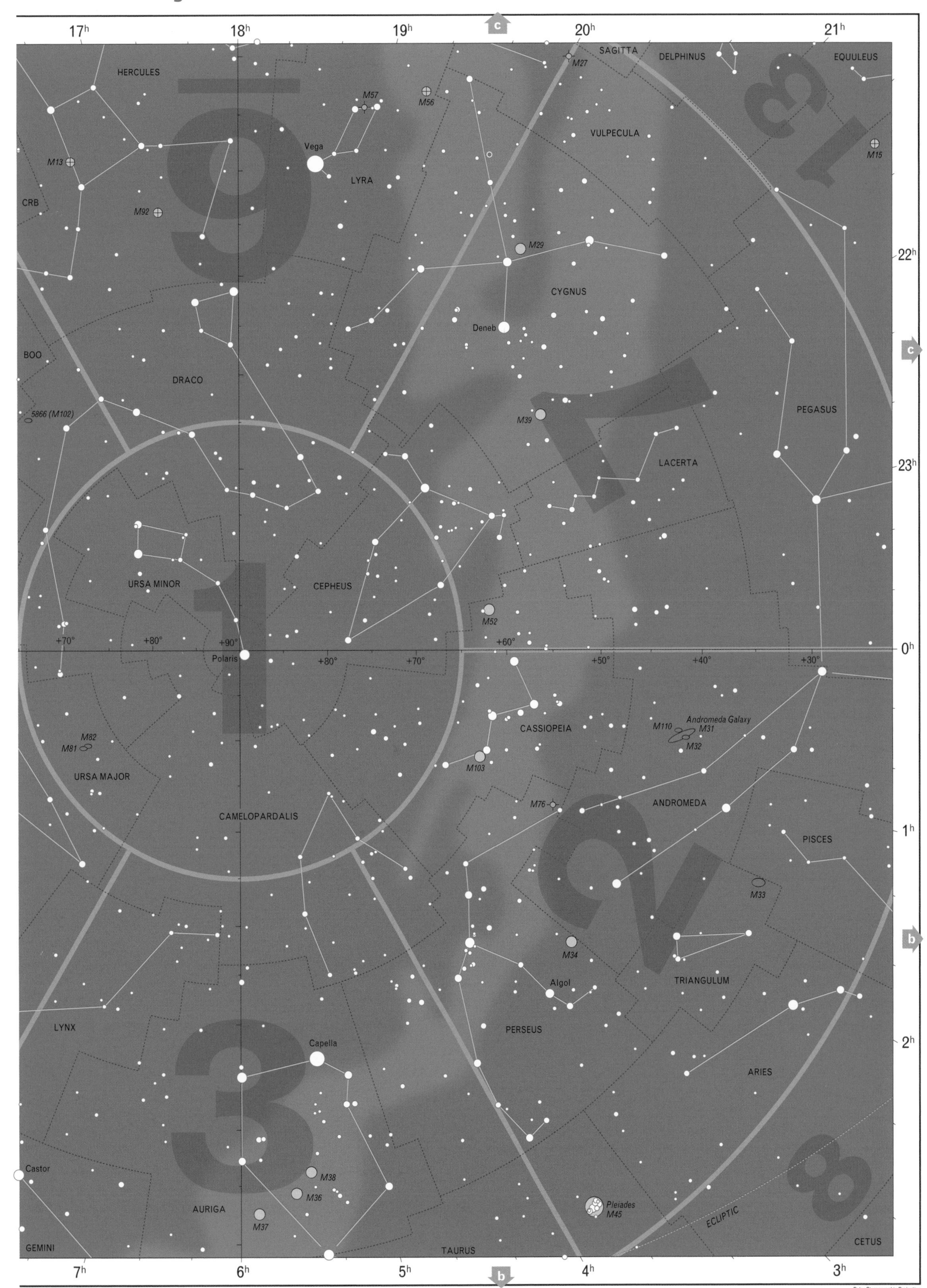


Polar Stereographic Projection

# Messier & Index Chart

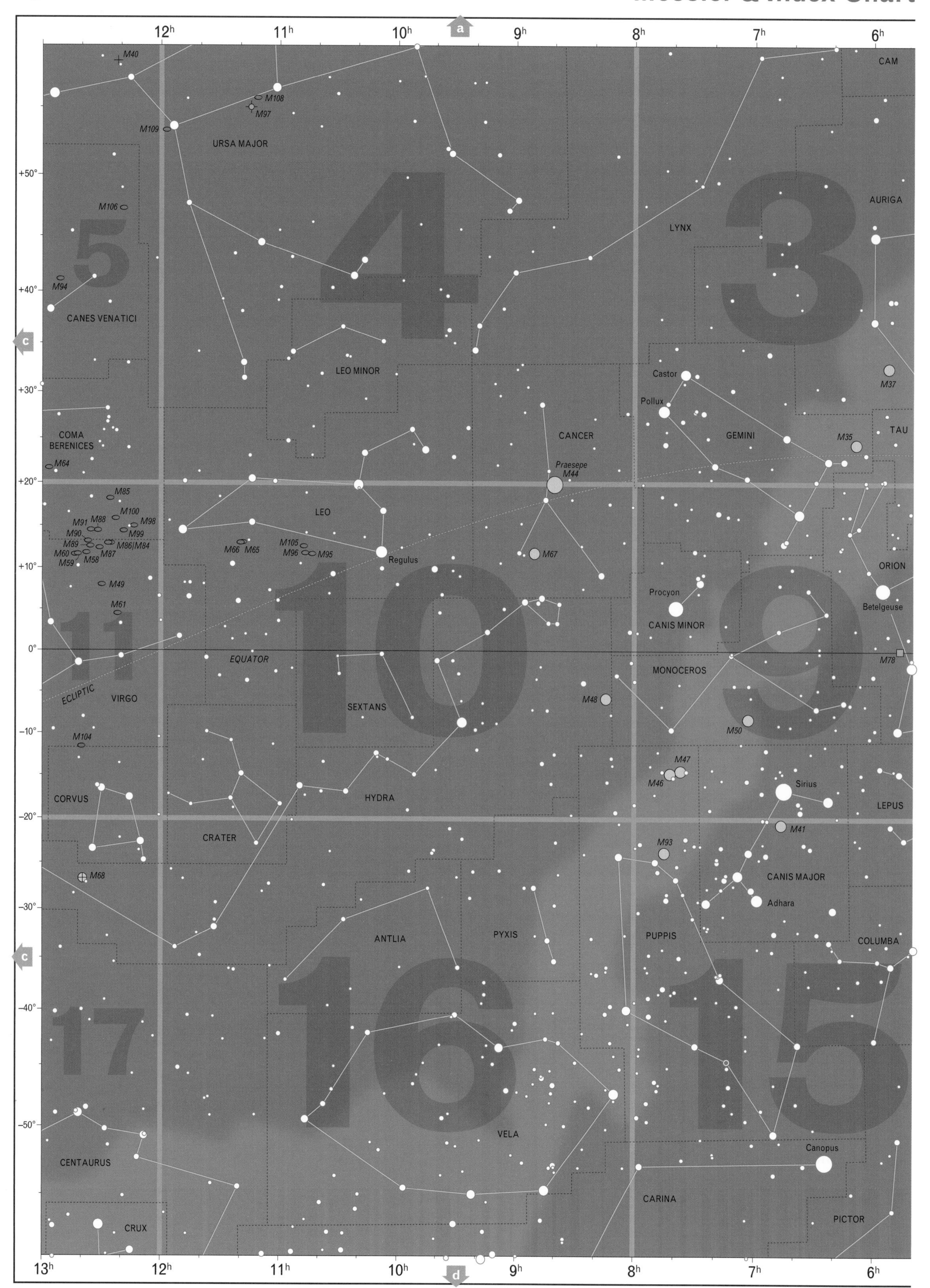

# Equatorial (RA $0^h - 12^h$)



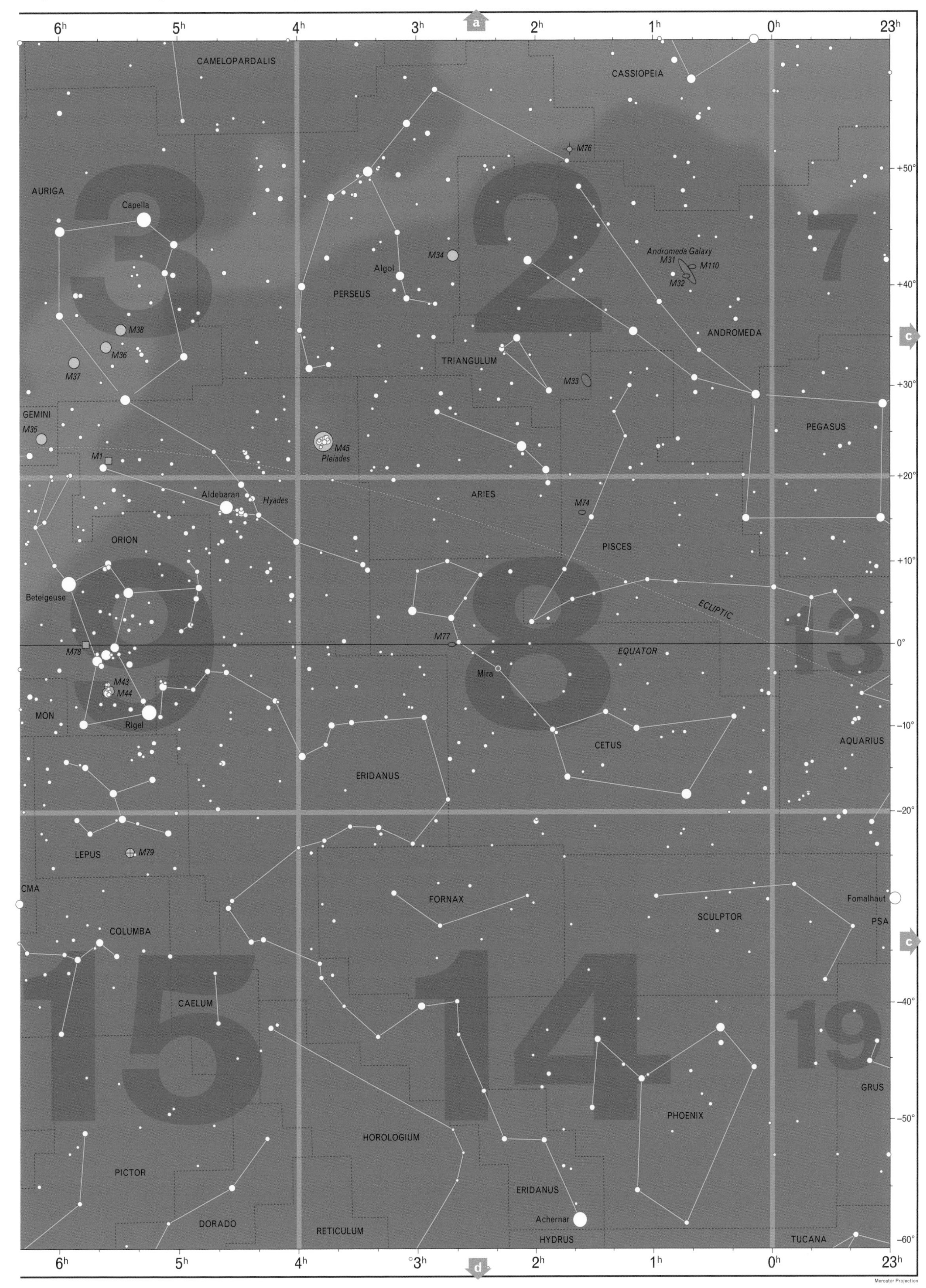

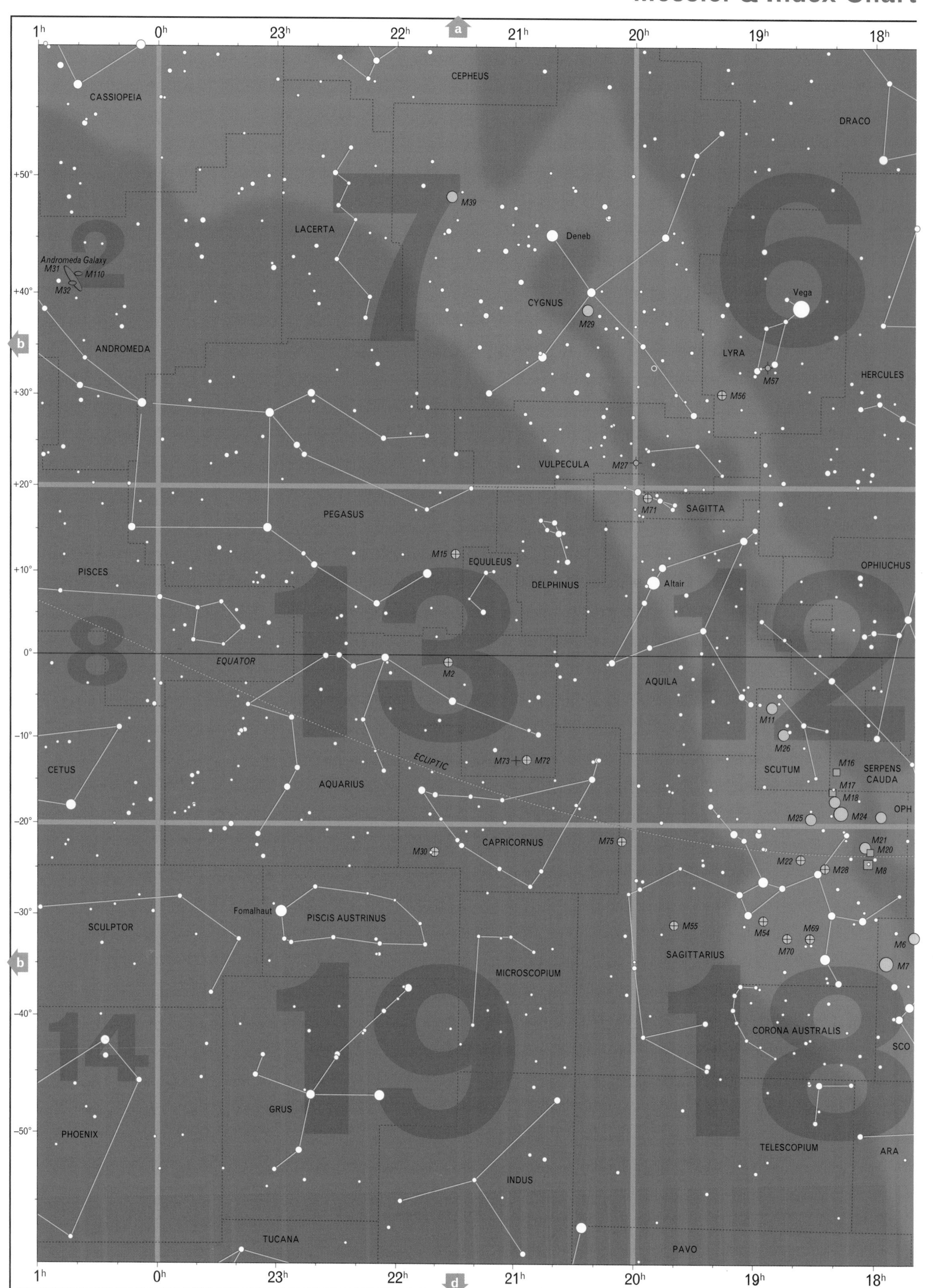
1h
0h
23h
22h
21h
20h
19h
18h
a
b
d
+50°
+40°
+30°
+20°
+10°
0°
-10°
-20°
-30°
-40°
-50°
CASSIOPEIA
CEPHEUS
DRACO
LACERTA
Deneb
CYGNUS
Vega
LYRA
HERCULES
Andromeda Galaxy
M31
M110
M32
M39
M29
M57
M56
ANDROMEDA
VULPECULA
M27
SAGITTA
M71
PEGASUS
M15
EQUULEUS
DELPHINUS
Altair
OPHIUCHUS
PISCES
EQUATOR
M2
AQUILA
M11
M26
SCUTUM
M16
SERPENS
CAUDA
M17
M18
M24
M25
OPH
CETUS
ECLIPTIC
M73
M72
AQUARIUS
CAPRICORNUS
M75
M30
M21
M20
M22
M28
M8
Fomalhaut
PISCIS AUSTRINUS
SCULPTOR
M55
M54
M69
M70
M6
M7
SAGITTARIUS
MICROSCOPIUM
CORONA AUSTRALIS
SCO
GRUS
PHOENIX
TELESCOPIUM
ARA
INDUS
TUCANA
PAVO
2
7
6
8
13
12
14
19
18

# Equatorial (RA $12^h - 0^h$)



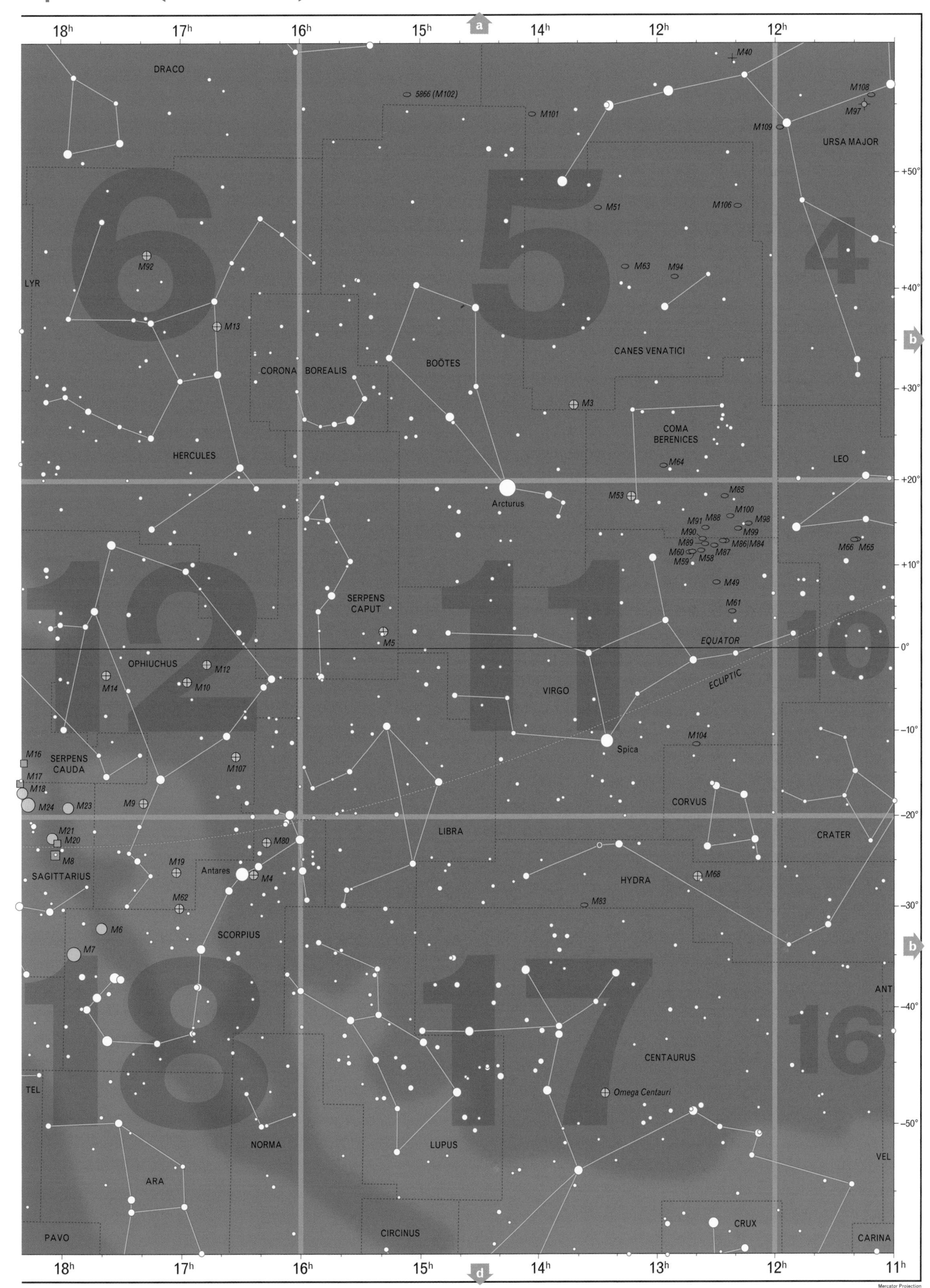

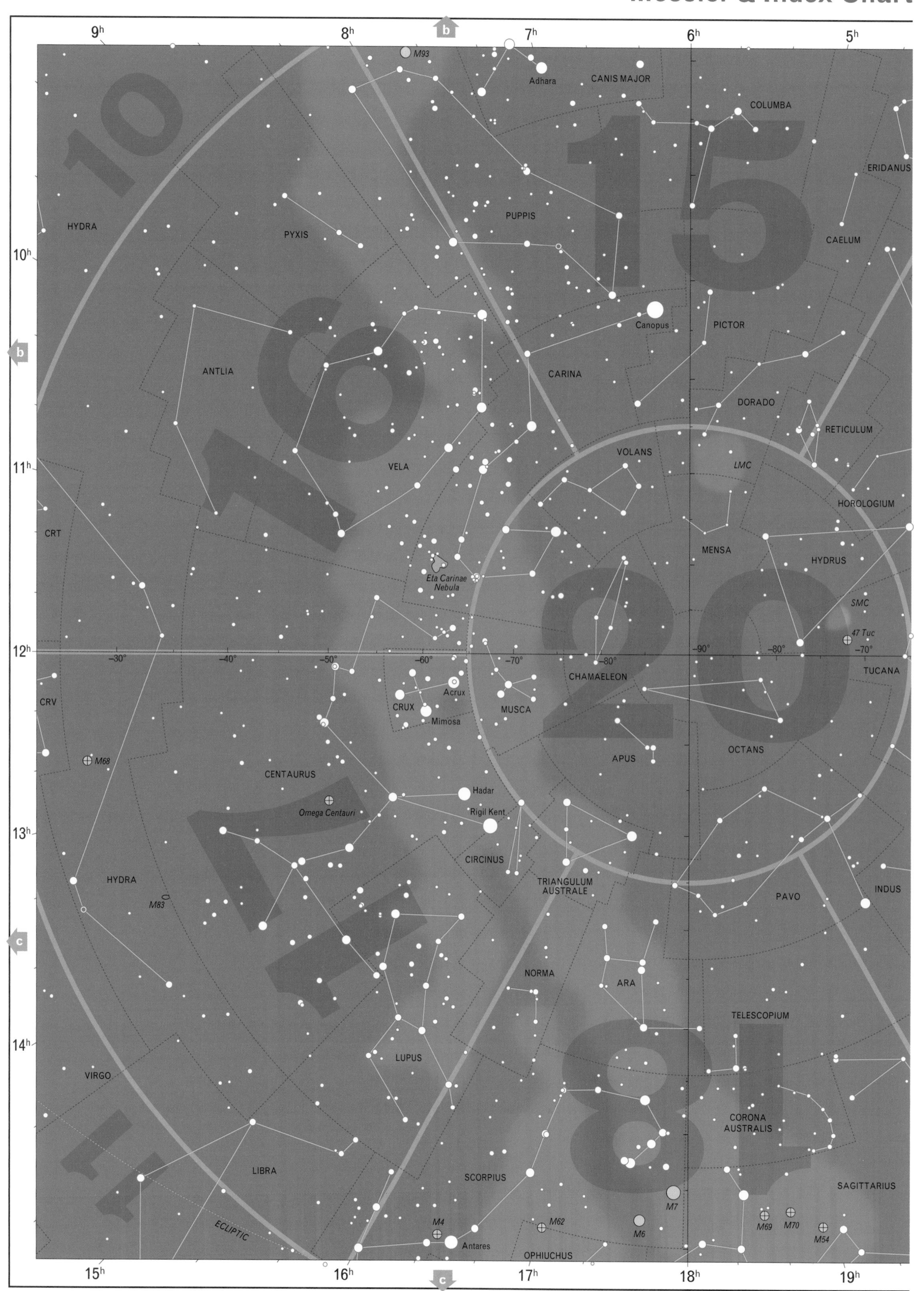
9h
8h
7h
6h
5h
b
M93
Adhara
CANIS MAJOR
COLUMBA
ERIDANUS
HYDRA
PYXIS
PUPPIS
CAELUM
10h
Canopus
PICTOR
ANTLIA
CARINA
DORADO
RETICULUM
VELA
VOLANS
LMC
11h
HOROLOGIUM
CRT
MENSA
HYDRUS
Eta Carinae Nebula
SMC
47 Tuc
12h
-30°
-40°
-50°
-60°
-70°
-80°
-90°
-80°
-70°
CHAMAELEON
TUCANA
CRV
Acrux
CRUX
MUSCA
Mimosa
M68
APUS
OCTANS
CENTAURUS
Hadar
Omega Centauri
Rigil Kent
13h
CIRCINUS
HYDRA
TRIANGULUM AUSTRALE
PAVO
INDUS
M83
NORMA
ARA
TELESCOPIUM
14h
LUPUS
VIRGO
CORONA AUSTRALIS
LIBRA
SCORPIUS
SAGITTARIUS
M7
ECLIPTIC
M4
M62
M6
M69
M70
M54
Antares
OPHIUCHUS
15h
16h
17h
18h
19h
c

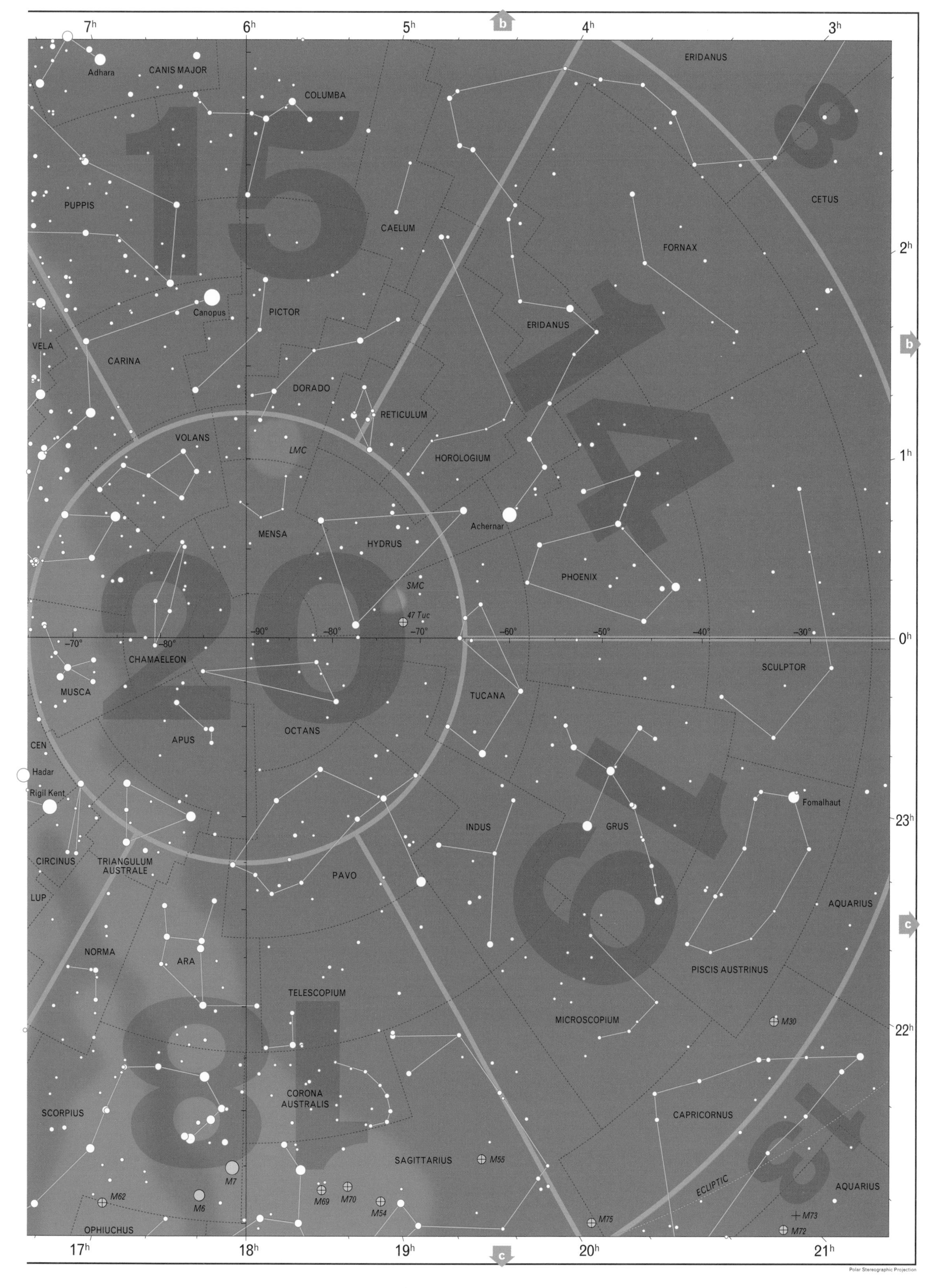

7h
6h
5h
4h
3h
b
Adhara
CANIS MAJOR
COLUMBA
ERIDANUS
PUPPIS
CAELUM
CETUS
FORNAX
2h
Canopus
PICTOR
ERIDANUS
b
VELA
CARINA
DORADO
RETICULUM
VOLANS
LMC
HOROLOGIUM
1h
MENSA
HYDRUS
Achernar
PHOENIX
SMC
47 Tuc
-90°
-80°
-70°
-60°
-50°
-40°
-30°
0h
CHAMAELEON
SCULPTOR
MUSCA
TUCANA
APUS
OCTANS
CEN
Hadar
Rigil Kent
Fomalhaut
INDUS
GRUS
23h
CIRCINUS
TRIANGULUM AUSTRALE
PAVO
LUP
AQUARIUS
c
NORMA
ARA
PISCIS AUSTRINUS
TELESCOPIUM
MICROSCOPIUM
M30
22h
CORONA AUSTRALIS
SCORPIUS
CAPRICORNUS
SAGITTARIUS
M55
M7
ECLIPTIC
AQUARIUS
M62
M6
M69
M70
M54
M75
M73
M72
OPHIUCHUS
17h
18h
19h
20h
21h
c
Polar Stereographic Projection

# STAR CHARTS

Once you are familiar with the sky and can recognize the different constellations, you will want to know more about what is visible in the night sky. So the next step is to use the star charts in this chapter, the 'heart' of *The Cambridge Star Atlas.* These star charts divide the sky into 20 parts. The actual chart areas are shown on the index to the star charts on page 39, preceding the star charts themselves and also on the Messier charts on pages 22–29. There is a generous overlap between the charts, so most of the constellations are shown complete on at least one chart. The positions of stars and objects are for the year 2000 or, to be more precise, the epoch is 2000.0 (the extra 0 is a decimal and it means January 1; for example 2000.5 would be July 1). The positions are plotted against a grid of right ascension (RA) and declination (Dec) comparable with longitude and latitude on the Earth's globe. *Right ascension* (RA) is reckoned in hours, minutes, and seconds from $0^h$ to $24^h$, and is measured from west to east along the Equator. *Declination* (Dec) represents the angular distance between an object and the celestial Equator, (+) for objects north and (–) for those south of the Equator (figure 5).

The charts' projections have been chosen carefully to show the star patterns with the least possible distortion and to make it easy to measure positions from the maps. The tick marks along the chart borders and on the central RA line (horizontal RA line on the polar maps) will also help you.

Figure 5

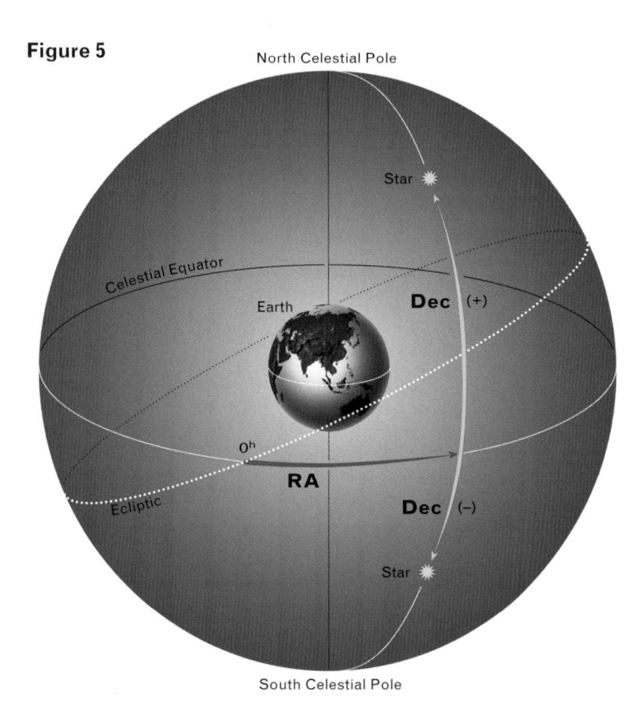


## Stars

Stars are huge balls of hot gas, like our own Sun, but much further away than the Sun. Although they appear as small points of light (so small that even in the largest telescopes most of them cannot be measured) many stars are even bigger than our Sun, which turns out to be of only average size. Some stars, like Betelgeuse (in Orion) and Antares (in Scorpius), have diameters that are close to half a billion kilometers (300 times that of our Sun). Even that is not the limit: some stars have diameters a thousand times larger than our Sun. At the other end of the scale are stars that are no bigger than the Earth, and some that are even smaller.

It is not only the sizes of stars that differ, but also their temperatures. The temperature of the Sun is almost 5,500 °C. That is just its surface temperature; in the nucleus of the Sun the temperature reaches millions of degrees. The Sun is a star that is radiating yellowish light. 'Cooler' stars are orange (like Aldebaran in Taurus, or Arcturus in Boötes, with surface temperatures of 4,000 °C) or red (Betelgeuse and Antares; 3,000 °C). But many stars are much hotter than the Sun. White stars have temperatures between 6,000 and 10,000 °C (Sirius in Canis Major, or Deneb in Cygnus). Then there are the bluish stars with temperatures between 11,000 and 40,000 °C (like Rigel in Orion). It is interesting to take a look at the beautiful constellation Orion on a dark night in December or January. You can clearly see the difference between the red giant star Betelgeuse and the bluish Rigel. Try it.

Stars are divided into spectral classes, according to differences in their spectra. These differences are related to the surface temperatures and the colors of the stars. The hottest, bluish stars are the classes O, B and A, the slightly cooler white stars are class F, yellow class G, orange class K, and the coolest, the red stars, are class M. To remember this rather illogical sequence of letters, use the sentence: **O**h, **B**e **A** **F**ine **G**irl, **K**iss **M**e! or if you prefer: **O**h, **B**e **A** **F**ine **G**uy, **K**iss **M**e! In the tables facing the star charts, the spectral classes of variable stars are given in the final column. The spectral types of the 96 brightest stars can be found in table C on the next two pages.

## Magnitudes

The word *magnitude* usually refers to the apparent brightness of a star or an object. Traditionally, stars visible to the naked eye were classified into six groups of brightness. The most prominent stars in the sky were called first magnitude stars, the ones slightly fainter (e.g. the Pole Star and the brightest stars of the Great Bear) second magnitude and so on. The faintest stars visible to the eye were magnitude six. Nowadays astronomers are able to measure the brightness very accurately. In modern catalogs you will find magnitude to two decimal places and the scale now has a logarithmic basis. A difference of five magnitudes is defined as being 100 times as bright. Consequently one magnitude represents a difference of 2.5 times (or to be precise 2.512 times, the fifth root of 100). On this scale several stars turn out to be brighter than 1, so the scale was extended to 0, but even that was not enough. A few stars are brighter still and are given a negative magnitude. The brightest star in the sky, Sirius, has a magnitude of -1.44, the second brightest, Canopus, is of magnitude -0.62.

On the seasonal sky maps (pages 12–19), the brightness of a star is rounded to the nearest whole magnitude. Stars between 0.5 and 1.5 are classified as magnitude 1, between 1.5 and 2.5 magnitude 2 and so on. For the star charts (pages 40–79), stars are classified to half magnitudes. A detailed scale can be found in the legend at the right of each chart. Stars' positions and magnitudes are based on the information from the *Hipparcos* and *Tycho Catalogues* (ESA Publications, 1997). All together approximately 9,500 stars are plotted. Furthermore, the Atlas shows 921 non-stellar objects: star clusters, nebulae, and distant galaxies.

Since stars are at very different distances, the apparent, or visual (V), magnitude does not show how bright the star really is compared with other stars or with our own Sun. Astronomers use the term *absolute magnitude* for the real brightness of a star ($M_V$). The absolute magnitude tells you how bright a star would appear in the sky if it was at a distance of 10 parsecs. One parsec equals a distance of 3.26 light-years. So 10 parsecs equals 32.6 light-years, or in other words the distance light travels in 32.6 years. That is $9.46 \times 10^{12}$ (9,460,000,000,000) kilometers. The speed of light is an amazing 299,792 kilometers per second.

In table C (pages 32–33) you will find both the *visual* (V) and the *absolute* ($M_V$) magnitudes, and a column telling you the distance in light-years. It is interesting to compare these columns. The brightest star in the sky (Sirius) is not really that bright when you compare its absolute magnitude (1.5) with that of Deneb (-7.5). This star shines about 1,500 times as bright as Sirius.

Although the Sun is not listed in the table, its absolute magnitude (4.8) shows that it is only a very average star, fainter than all the stars in the list, although it is obviously the brightest object in our sky, with a visual magnitude of -26.8. It shines so brightly because on the cosmic scale it is very close, 'only' 8 light-minutes! When you compare Deneb with our Sun you will find that it is almost 100,000 times as bright as the Sun.

A lower case letter v after an entry in the visual magnitude column in table C means that this is a so-called *variable star*. This is explained in section 'Variable stars' on page 34.

## Stars names

Many of the brighter stars have proper names of Latin, Greek, or Arabic origin, such as Regulus, Aldebaran, and Betelgeuse. Only the brightest stars are still referred to by these old names. On the star

*Continued on page 34*

**Table C** *The 96 brightest stars*

**All stars brighter than magnitude 2.56 (V), in order of brightness**

| Star | Name | RA | | Dec | | Magnitudes | | Spectral | Distance | Chart |
|---|---|---|---|---|---|---|---|---|---|---|
| | | h | m | ° | ′ | V | $M_v$ | type | (light-years) | numbers |
| α CMa | Sirius | 06 | 45.2 | −16 | 43 | −1.44 | 1.5 | A0 | 9 | 9 15 |
| α Car | Canopus | 06 | 23.9 | −52 | 41 | −0.62 | −5.4 | F0 | 313 | 14 15 16 |
| α Cen | Rigil Kent | 14 | 39.6 | −60 | 50 | −0.29 | 4.0 | G2 + K1 | 4 | 17 18 20 |
| α Boo | Arcturus | 14 | 15.7 | +19 | 11 | −0.05 | −0.6 | K2 | 37 | 5 11 |
| α Lyr | Vega | 18 | 36.9 | +38 | 47 | 0.03 | 0.6 | A0 | 25 | 6 7 |
| α Aur | Capella | 05 | 16.7 | +46 | 00 | 0.08 | −0.8 | G6 | 42 | 2 3 |
| β Ori | Rigel | 05 | 14.5 | −08 | 12 | 0.18 | −6.6 | B8 | 773 | 9 |
| α CMi | Procyon | 07 | 39.3 | +05 | 14 | 0.40 | 2.8 | F5 | 11 | 9 10 |
| α Ori | Betelgeuse | 05 | 55.2 | +07 | 24 | 0.45 v | −5.0 | M2 | 427 | 9 |
| α Eri | Achernar | 01 | 37.6 | −57 | 14 | 0.45 | −2.9 | B3 | 144 | 14 15 19 20 |
| β Cen | Hadar | 14 | 03.8 | −60 | 22 | 0.61 v | −5.5 | B1 | 525 | 16 17 18 20 |
| α Aql | Altair | 19 | 50.8 | +08 | 52 | 0.76 | 2.1 | A7 | 17 | 7 12 13 |
| α Cru | Acrux | 12 | 26.6 | −63 | 06 | 0.77 | −4.6 | B0 | 321 | 16 17 20 |
| α Tau | Aldebaran | 04 | 35.9 | +16 | 31 | 0.87 v | −0.8 | K5 | 65 | 2 3 8 9 |
| α Vir | Spica | 13 | 25.2 | −11 | 10 | 0.98 v | −3.6 | B1 | 262 | 11 |
| α Sco | Antares | 16 | 29.4 | −26 | 26 | 1.06 v | −5.8 | M1 | 604 | 11 12 17 18 |
| β Gem | Pollux | 07 | 45.3 | +28 | 02 | 1.16 | 1.1 | K0 | 34 | 3 4 9 10 |
| α PsA | Fomalhaut | 22 | 57.7 | −29 | 37 | 1.17 | 1.6 | A3 | 25 | 14 19 |
| α Cyg | Deneb | 20 | 41.4 | +45 | 17 | 1.25 | −7.5 | A2 | 3,262 | 6 7 |
| β Cru | Mimosa | 12 | 47.7 | −59 | 41 | 1.25 v | −4.0 | B0 | 353 | 16 17 20 |
| α Leo | Regulus | 10 | 08.4 | +11 | 58 | 1.36 | −0.6 | B7 | 77 | 10 |
| ε CMa | Adhara | 06 | 58.6 | −28 | 58 | 1.50 | −4.1 | B2 | 431 | 9 15 16 |
| α Gem | Castor | 07 | 34.6 | +31 | 53 | 1.58 | 0.6 | A2 | 52 | 3 4 |
| γ Cru | Gacrux | 12 | 31.2 | −57 | 07 | 1.59 v | −0.8 | M4 | 88 | 16 17 20 |
| λ Sco | Shaula | 17 | 33.6 | −37 | 06 | 1.62 v | −3.6 | B1 | 701 | 17 18 |
| γ Ori | Bellatrix | 05 | 25.1 | +06 | 21 | 1.64 | −2.8 | B2 | 243 | 9 |
| β Tau | Alnath | 05 | 26.3 | +28 | 36 | 1.65 | −1.3 | B7 | 131 | 3 9 |
| β Car | Miaplacides | 09 | 13.2 | −69 | 43 | 1.67 | −1.1 | A2 | 111 | 16 20 |
| ε Ori | Alnilam | 05 | 36.2 | −01 | 12 | 1.69 | −6.6 | B0 | 1,345 | 9 |
| α Gru | Alnair | 22 | 08.2 | −46 | 58 | 1.73 | −0.9 | B7 | 101 | 14 18 19 20 |
| ζ Ori | Alnitak | 05 | 40.8 | −01 | 57 | 1.74 | −5.5 | O9 | 816 | 9 |
| γ Vel | — | 08 | 09.5 | −47 | 20 | 1.75 v | −5.8 | WC8 + O9 | 842 | 15 16 |
| ε UMa | Alioth | 12 | 54.0 | +55 | 58 | 1.76 v | −0.2 | A0 | 81 | 1 4 5 |
| α Per | Mirphak | 03 | 24.3 | +49 | 52 | 1.79 | −4.9 | F5 | 593 | 2 3 |
| ε Sgr | Kaus Australis | 18 | 24.2 | −34 | 23 | 1.79 | −1.4 | B9 | 145 | 18 |
| α UMa | Dubhe | 11 | 03.7 | +61 | 45 | 1.81 | −1.3 | F7 | 124 | 1 4 5 |
| δ CMa | Wezen | 07 | 08.4 | −26 | 23 | 1.83 | −7.2 | F8 | 1,787 | 9 15 16 |
| η UMa | Alkaid | 13 | 47.5 | +49 | 19 | 1.85 | −1.8 | B3 | 101 | 1 4 5 6 |
| ε Car | Avior | 08 | 22.5 | −59 | 31 | 1.86 | −4.8 | K3 + B2 | 633 | 15 16 20 |
| ϑ Sco | — | 17 | 37.3 | −43 | 00 | 1.86 | −3.0 | F1 | 272 | 17 18 |
| β Aur | Menkalinan | 05 | 59.5 | +44 | 57 | 1.90 v | −0.2 | A2 | 82 | 2 3 4 |
| α TrA | Atria | 16 | 48.7 | −69 | 02 | 1.91 | −5.0 | K2 | 415 | 17 18 20 |
| δ Vel | — | 08 | 44.7 | −54 | 43 | 1.93 | 0.0 | A1 | 80 | 15 16 20 |
| γ Gem | Alhena | 06 | 37.7 | +16 | 24 | 1.93 | −0.6 | A0 | 105 | 3 9 |
| α Pav | Peacock | 20 | 25.6 | −56 | 44 | 1.94 | −2.1 | B2 | 183 | 18 19 20 |
| α UMi | Polaris | 02 | 31.8 | +89 | 16 | 1.97 v | −4.1 | F7 | 432 | 1 |
| β CMa | Mirzam | 06 | 22.7 | −17 | 57 | 1.98 v | −4.0 | B1 | 500 | 9 15 |
| α Hya | Alphard | 09 | 27.6 | −08 | 39 | 1.99 | −2.1 | K3 | 177 | 10 |

**Table C** *The 96 brightest stars (continued)*

**All stars brighter than magnitude 2.56 (V), in order of brightness**

| Star | Name | RA | | Dec | | Magnitudes | | Spectral type | Distance (light-years) | Chart numbers |
|---|---|---|---|---|---|---|---|---|---|---|
| | | h | m | ° | ′ | V | $M_v$ | | | |
| γ Leo | Algieba | 10 | 20.0 | +19 | 50 | 2.01 | −2.2 | K0 | 126 | 4 10 |
| α Ari | Hamal | 02 | 07.2 | +23 | 28 | 2.01 | 0.5 | K2 | 66 | 2 8 |
| β Cet | Deneb Kaitos | 00 | 43.6 | −17 | 59 | 2.04 | −1.0 | K0 | 96 | 8 14 |
| σ Sgr | Nunki | 18 | 55.3 | −26 | 18 | 2.05 | −2.4 | B2.5 | 224 | 12 18 |
| β Gru | — | 22 | 42.7 | −46 | 53 | 2.06 v | −1.4 | M5 | 170 | 14 19 |
| ϑ Cen | Menkent | 14 | 06.7 | −36 | 22 | 2.06 | 0.1 | K0 | 61 | 17 |
| β UMi | Kochab | 14 | 50.7 | +74 | 09 | 2.07 | −1.1 | K4 | 126 | 1 |
| α And | Alpheratz | 00 | 08.4 | +29 | 05 | 2.07 | −0.6 | B9 | 97 | 2 7 8 13 |
| β And | Mirach | 01 | 09.7 | +35 | 37 | 2.07 | −1.9 | M0 | 199 | 2 7 |
| κ Ori | Saiph | 05 | 47.8 | −09 | 40 | 2.07 | −5.0 | B0 | 721 | 9 |
| α Oph | Rasalhague | 17 | 34.9 | +12 | 34 | 2.08 | 1.3 | A5 | 47 | 12 |
| β Per | Algol | 03 | 08.2 | +40 | 57 | 2.09 v | −0.5 | B8 | 93 | 2 3 |
| γ And | Almaak | 02 | 03.9 | +42 | 20 | 2.10 | −3.0 | B8 | 355 | 2 3 7 |
| β Leo | Denebola | 11 | 49.1 | +14 | 34 | 2.14 | 1.9 | A3 | 36 | 4 5 10 11 |
| γ Cas | — | 00 | 56.7 | +60 | 43 | 2.15 v | −5.0 | B0 | 613 | 1 2 7 |
| γ Cen | — | 12 | 41.5 | −48 | 58 | 2.20 | −0.6 | A1 | 130 | 16 17 |
| ζ Pup | — | 08 | 03.6 | −40 | 00 | 2.21 | −6.1 | O5 | 1,403 | 15 16 |
| ι Car | Aspidiske | 09 | 17.1 | −59 | 17 | 2.21 v | −4.4 | A8 | 694 | 15 16 20 |
| α CrB | Alphekka | 15 | 34.7 | +26 | 43 | 2.22 v | 0.3 | A0 | 75 | 5 6 11 12 |
| λ Vel | — | 09 | 08.0 | −43 | 26 | 2.23 v | −4.8 | K4 | 572 | 15 16 |
| γ Cyg | Sadr | 20 | 22.2 | +40 | 15 | 2.23 | −4.1 | F8 | 1,517 | 6 7 |
| ζ UMa | Mizar | 13 | 23.9 | +54 | 56 | 2.23 | 0.3 | A2 | 78 | 1 4 5 6 |
| γ Dra | Etamin | 17 | 56.6 | +51 | 29 | 2.24 | −1.1 | K5 | 148 | 5 6 7 |
| α Cas | Schedir | 00 | 40.5 | +56 | 32 | 2.24 | −2.5 | K0 | 229 | 1 2 7 |
| δ Ori | Mintaka | 05 | 32.0 | −00 | 18 | 2.25 v | −0.6 | O9 | 919 | 9 |
| β Cas | Caph | 00 | 09.2 | +59 | 09 | 2.27 v | 1.2 | F2 | 54 | 1 2 7 |
| ε Sco | — | 16 | 50.2 | −34 | 18 | 2.29 | 0.1 | K2 | 65 | 17 18 |
| α Lup | — | 14 | 41.9 | −47 | 23 | 2.29 v | −4.1 | B1 | 548 | 17 18 20 |
| ε Cen | — | 13 | 39.9 | −53 | 28 | 2.29 v | −3.3 | B1 | 376 | 16 17 18 20 |
| δ Sco | — | 16 | 00.3 | −22 | 37 | 2.29 | −4.4 | B0 | 401 | 11 12 17 18 |
| η Cen | — | 14 | 35.5 | −42 | 10 | 2.33 v | −2.8 | B1 | 309 | 17 18 |
| β UMa | Merak | 11 | 01.8 | +56 | 23 | 2.34 | 0.4 | A1 | 79 | 1 3 4 5 |
| ε Boo | Izar | 14 | 45.0 | +27 | 04 | 2.35 | −2.6 | A0 | 210 | 5 11 |
| ε Peg | Enif | 21 | 44.2 | +09 | 52 | 2.38 v | −5.2 | K2 | 672 | 13 |
| κ Sco | — | 17 | 42.5 | −39 | 02 | 2.39 v | −3.6 | B1 | 464 | 17 18 |
| α Phe | Ankaa | 00 | 26.3 | −42 | 18 | 2.40 | −0.3 | K0 | 77 | 14 19 |
| γ UMa | Phad | 11 | 53.8 | +53 | 42 | 2.41 | 0.2 | A0 | 84 | 1 4 5 |
| η Oph | Sabik | 17 | 10.4 | −15 | 43 | 2.43 | 0.8 | A2 | 84 | 12 18 |
| β Peg | Scheat | 23 | 03.8 | +28 | 05 | 2.44 v | −1.7 | M2 | 199 | 2 7 13 |
| α Cep | Alderamin | 21 | 18.6 | +62 | 35 | 2.45 | 1.4 | A7 | 49 | 1 6 7 |
| η CMa | Aludra | 07 | 24.1 | −29 | 18 | 2.45 | −7.5 | B5 | 3,182 | 9 15 16 |
| κ Vel | — | 09 | 22.1 | −55 | 00 | 2.47 | −3.9 | B2 | 539 | 15 16 17 20 |
| ε Cyg | — | 20 | 46.2 | +33 | 58 | 2.48 | 0.7 | K0 | 72 | 6 7 |
| α Peg | Merkab | 23 | 04.8 | +15 | 12 | 2.49 | −0.9 | B9 | 140 | 7 13 |
| α Cet | Menkar | 03 | 02.3 | +04 | 05 | 2.54 | −1.7 | M2 | 220 | 8 |
| ζ Cen | — | 13 | 55.5 | −47 | 17 | 2.54 | −2.9 | B2 | 385 | 16 17 18 20 |
| ζ Oph | — | 16 | 37.2 | −10 | 34 | 2.54 | −4.3 | O9 | 458 | 11 12 |
| β Sco | Graffias | 16 | 05.4 | −19 | 48 | 2.56 | −3.5 | B1 + B2 | 530 | 11 12 17 18 |

**Table D** *The Greek alphabet*

| | | | | | |
|---|---|---|---|---|---|
| α | Alpha | ι | Iota | ρ | Rho |
| β | Beta | κ | Kappa | σ (ς) | Sigma |
| γ | Gamma | λ | Lambda | τ | Tau |
| δ | Delta | μ | Mu | υ | Upsilon |
| ε | Epsilon | ν | Nu | ϕ (φ) | Phi |
| ζ | Zeta | ξ | Xi | χ | Chi |
| η | Eta | ο | Omicron | ψ | Psi |
| θ (ϑ) | Theta | π | Pi | ω | Omega |

charts you will find all proper names that are listed in table C. Most stars also bear designations of numbers and Greek letters. The German celestial cartographer Johann Bayer introduced the Greek letters in the beginning of the seventeenth century. In general, the brightest star in a constellation was given the first letter of the Greek alphabet, alpha (α). The second letter, beta (β), was assigned to the second brightest and so on, although there are many obvious exceptions to this rule. In table C you will find the Greek letters and their names.

Another way to identify stars is by numbers. In each constellation the stars are numbered in order of RA. These numbers are usually referred to as Flamsteed numbers. Most of the brighter stars have both a Greek letter and a Flamsteed number. On the star charts you will find all the Greek lettered stars and, in addition, the Flamsteed numbers for the other stars. Generally, when a star is referred to in the text or in a table by a Greek letter or by a Flamsteed number, the genitive form of the Latin constellation name, or its official abbreviation, follows it. So the star Deneb, in the constellation the Swan (Latin: *Cygnus*), can also be called Alpha Cygni or 50 Cygni (α Cyg, 50 Cyg). Table E, on page 35, gives the names of the constellations, the genitive form, the official abbreviation, and the common English names.

Variable stars (see next column) have a quite different nomenclature. Some have the regular star identification like Algol (β Per) or Mira (ο Cet) but most are labeled in a special way: by roman letters starting with R, then S, T etc. to Z. Then RR, RS, RT, . . . , RZ, next SS, ST, and so on up to ZZ. After these 54 the naming continues with AA, AB, . . . , AZ, then BB, BC and so on again up to QZ, giving a total 334. The next variable found in that constellation is called V335, then V336 and so on. These designations are also followed by the constellation name, as with the Greek letters and the Flamsteed numbers.

## Constellations

Most of the constellations we know originate from Mesopotamian traditions and from Greek mythology, but over the centuries many other constellations have been added to the classical ones, especially in the southern sky. Several of these new constellations only had a short life, while others have survived. In 1930 the International Astronomical Union (IAU) finally adopted a list of 88 official constellations and the boundaries were also delimited once and for all. On our star charts these official boundaries are drawn as purple dashed lines. The 88 official constellations are listed in table E. The final column of the table gives the numbers of the charts where you can best find each of these constellations.

## Variable stars

The brightness of many stars varies over longer or shorter periods of time. The most common reason for this is that the size of the star actually changes; the star pulsates. A well-known type of pulsating star is the Cepheid, named after Delta (δ) Cephei, a yellow supergiant, which regularly pulsates every few days or weeks. The Cepheids are divided into two classes: the classical (Cep) and the Population II (CW) Cepheids. They are important to astronomers because there is a relation between their period and luminosity. The brighter a Cepheid, the longer the period. When the period is measured, we know the real luminosity (*absolute* brightness) of the star. By comparing this with the amount of light we actually receive (apparent or *visual* brightness) we have an important tool for calculating the star's distance.

Another type of pulsating variable is named after the prototype Omicron (ο) Ceti, or Mira (M), a red giant. These variables do not have a strict period. Several other types of pulsating variables are also named after prototypes, like U Gem, R CrB, or RR Lyr.

*Continued on page 36*

**Table E** *List of constellations*

| | Name | Genitive | Abbreviation | Common name | Chart numbers | | |
|---|---|---|---|---|---|---|---|
| 1 | **Andromeda** | Andromedae | And | *Andromeda* | 2 | | |
| 2 | **Antlia** | Antliae | Ant | *Air Pump* | 16 | | |
| 3 | **Apus** | Apodis | Aps | *Bird of Paradise* | 20 | | |
| 4 | **Aquarius** | Aquarii | Aqr | *Water Carrier* | 13 | | |
| 5 | **Aquila** | Aquilae | Aql | *Eagle* | 12 | | |
| 6 | **Ara** | Arae | Ara | *Altar* | 18 | | |
| 7 | **Aries** | Arietis | Ari | *Ram* | 8 | 2 | |
| 8 | **Auriga** | Aurigae | Aur | *Charioteer* | 3 | | |
| 9 | **Boötes** | Boötis | Boo | *Herdsman* | 5 | 11 | |
| 10 | **Caelum** | Caeli | Cae | *Engraving Tool* | 15 | | |
| 11 | **Camelopardalis** | Camelopardalis | Cam | *Giraffe* | 1 | 3 | |
| 12 | **Cancer** | Cancri | Cnc | *Crab* | 10 | 4 | |
| 13 | **Canes Venatici** | Canum Venaticorum | CVn | *Hunting Dogs* | 5 | | |
| 14 | **Canis Major** | Canis Majoris | CMa | *Greater Dog* | 9 | 15 | |
| 15 | **Canis Minor** | Canis Minoris | CMi | *Lesser Dog* | 9 | | |
| 16 | **Capricornus** | Capricorni | Cap | *Sea Goat* | 13 | | |
| 17 | **Carina** | Carinae | Car | *Keel* | 16 | 20 | |
| 18 | **Cassiopeia** | Cassiopeiae | Cas | *Cassiopeia* | 1 | 2 | |
| 19 | **Centaurus** | Centauri | Cen | *Centaur* | 17 | | |
| 20 | **Cepheus** | Cephei | Cep | *Cepheus* | 1 | | |
| 21 | **Cetus** | Ceti | Cet | *Whale* | 8 | | |
| 22 | **Chamaeleon** | Chamaeleonis | Cha | *Chameleon* | 20 | | |
| 23 | **Circinus** | Circini | Cir | *Pair of Compasses* | 17 | 20 | |
| 24 | **Columba** | Columbae | Col | *Dove* | 15 | | |
| 25 | **Coma Berenices** | Coma Berenicis | Com | *Berenice's Hair* | 5 | 11 | |
| 26 | **Corona Australis** | Coronae Australis | CrA | *Southern Crown* | 18 | | |
| 27 | **Corona Borealis** | Coronae Borealis | CrB | *Northern Crown* | 5 | 6 | |
| 28 | **Corvus** | Corvi | Crv | *Crow* | 11 | | |
| 29 | **Crater** | Crateris | Crt | *Cup* | 10 | | |
| 30 | **Crux** | Crucis | Cru | *Southern Cross* | 16 | 17 | 20 |
| 31 | **Cygnus** | Cygni | Cyg | *Swan* | 7 | | |
| 32 | **Delphinus** | Delphini | Del | *Dolphin* | 13 | | |
| 33 | **Dorado** | Doradus | Dor | *Goldfish* | 15 | | |
| 34 | **Draco** | Draconis | Dra | *Dragon* | 1 | 6 | |
| 35 | **Equuleus** | Equulei | Equ | *Little Horse* | 13 | | |
| 36 | **Eridanus** | Eridani | Eri | *River Eridanus* | 8 | 9 | 14 |
| 37 | **Fornax** | Fornacis | For | *Furnace* | 14 | | |
| 38 | **Gemini** | Geminorum | Gem | *Twins* | 3 | 9 | |
| 39 | **Grus** | Gruis | Gru | *Crane* | 19 | | |
| 40 | **Hercules** | Herculis | Her | *Hercules* | 6 | 12 | |
| 41 | **Horologium** | Horologii | Hor | *Pendulum Clock* | 14 | | |
| 42 | **Hydra** | Hydrae | Hya | *Water Snake* | 10 | 16 | 17 |
| 43 | **Hydrus** | Hydri | Hyi | *Lesser Water Snake* | 20 | | |
| 44 | **Indus** | Indi | Ind | *Indian* | 19 | 20 | |
| 45 | **Lacerta** | Lacertae | Lac | *Lizard* | 7 | | |
| 46 | **Leo** | Leonis | Leo | *Lion* | 10 | 4 | |
| 47 | **Leo Minor** | Leonis Minoris | LMi | *Lesser Lion* | 4 | | |
| 48 | **Lepus** | Leporis | Lep | *Hare* | 9 | | |
| 49 | **Libra** | Librae | Lib | *Scales* | 11 | 17 | |
| 50 | **Lupus** | Lupi | Lup | *Wolf* | 17 | 18 | |
| 51 | **Lynx** | Lyncis | Lyn | *Lynx* | 4 | | |
| 52 | **Lyra** | Lyrae | Lyr | *Lyre* | 6 | | |
| 53 | **Mensa** | Mensae | Men | *Table Mountain* | 20 | | |
| 54 | **Microscopium** | Microscopii | Mic | *Microscope* | 19 | | |
| 55 | **Monoceros** | Monocerotis | Mon | *Unicorn* | 9 | | |
| 56 | **Musca** | Muscae | Mus | *Fly* | 20 | | |
| 57 | **Norma** | Normae | Nor | *Level* | 17 | 18 | |
| 58 | **Octans** | Octantis | Oct | *Octant* | 20 | | |
| 59 | **Ophiuchus** | Ophiuchi | Oph | *Serpent Holder* | 12 | 18 | |
| 60 | **Orion** | Orionis | Ori | *Orion, the Hunter* | 9 | | |
| 61 | **Pavo** | Pavonis | Pav | *Peacock* | 20 | 18 | |
| 62 | **Pegasus** | Pegasi | Peg | *Pegasus* | 13 | 7 | |
| 63 | **Perseus** | Persei | Per | *Perseus* | 2 | 3 | |
| 64 | **Phoenix** | Phoenicis | Phe | *Phoenix* | 14 | | |
| 65 | **Pictor** | Pictoris | Pic | *Painter's Easel* | 15 | | |
| 66 | **Pisces** | Piscium | Psc | *Fishes* | 8 | 13 | 2 |
| 67 | **Piscis Austrinus** | Piscis Austrini | PsA | *Southern Fish* | 19 | | |
| 68 | **Puppis** | Puppis | Pup | *Stern* | 15 | 9 | |
| 69 | **Pyxis** | Pyxidis | Pyx | *Mariner's Compass* | 16 | | |
| 70 | **Reticulum** | Reticuli | Ret | *Net* | 14 | 15 | |
| 71 | **Sagitta** | Sagittae | Sge | *Arrow* | 6 | 12 | |
| 72 | **Sagittarius** | Sagittarii | Sgr | *Archer* | 18 | 12 | |
| 73 | **Scorpius** | Scorpii | Sco | *Scorpion* | 18 | 12 | |
| 74 | **Sculptor** | Sculptoris | Scl | *Sculptor* | 14 | 19 | |
| 75 | **Scutum** | Scuti | Sct | *Shield* | 12 | | |
| 76 | **Serpens** | Serpentis | Ser | *Serpent* | 12 | 11 | |
| 77 | **Sextans** | Sextantis | Sex | *Sextant* | 10 | | |
| 78 | **Taurus** | Tauri | Tau | *Bull* | 9 | 3 | |
| 79 | **Telescopium** | Telescopii | Tel | *Telescope* | 18 | 19 | |
| 80 | **Triangulum** | Trianguli | Tri | *Triangle* | 2 | | |
| 81 | **Triangulum Australe** | Trianguli Australis | TrA | *Southern Triangle* | 20 | | |
| 82 | **Tucana** | Tucanae | Tuc | *Toucan* | 20 | | |
| 83 | **Ursa Major** | Ursae Majoris | UMa | *Great Bear* | 4 | 1 | |
| 84 | **Ursa Minor** | Ursae Minoris | UMi | *Lesser Bear* | 1 | | |
| 85 | **Vela** | Velorum | Vel | *Sail* | 16 | | |
| 86 | **Virgo** | Virginis | Vir | *Virgin* | 11 | | |
| 87 | **Volans** | Volantis | Vol | *Flying Fish* | 20 | | |
| 88 | **Vulpecula** | Vulpeculae | Vul | *Fox* | 6 | 7 | |

There is also a completely different type of variable star: the eclipsing variable. An eclipsing variable star is a double star in mutual orbit and one component periodically moves in front of or behind the other, causing a drop in the total amount of light we receive. The best-known example of this type is Beta (β) Persei (Algol). Eclipsing variables are referred to in the star chart tables as type E, subdivided by the shape of their light curves into E, EA, EB, and EW.

A further group is termed the eruptive variables. These undergo a very sudden and very large increase in brightness. The best known of these are the novae (N) and the supernovae (SN). A nova is a very close double star, in which one component is a white dwarf: a small but very compact star. Gas from the other component flows into the white dwarf and eventually it ignites in a huge explosion. The brightness of the star increases temporarily by thousands of times. Some novae erupt more than once. These are known as recurrent novae (Rn).

A supernova is even more spectacular. It is the catastrophic death of a very hot star. The star's life ends when it blows itself up and for a short period it shines millions of times brighter than it once did. After the star has faded again the outer shells of the star form a slowly expanding nebula. The Crab Nebula (M1) in Taurus (the Bull) is one example. In the star charts all variable stars with a maximum of magnitude 6.5 or brighter are plotted.

## Double stars

The majority of stars are double or multiple stars. Sometimes two stars appear very close in the sky, but are really only on the same line of sight; their distances differ considerably. These are called optical double stars. Real physical double stars belong together and are also called binaries. They are tied together by gravity and are in mutual orbit. The same goes for triple, quadruple, and even larger families of stars. Their apparent separation is measured in minutes and seconds of arc. One degree (°) can be divided into 60 minutes (′) and one minute divided again into 60 seconds (″). The separation is given in the tables (column with heading Sep) in seconds of arc. There is also a column with the heading PA, meaning position angle. This gives the angular position of the fainter component in relation to the brighter one. The angle is measured from the north, eastward. Bear in mind that in the sky east and west are reversed. So, when north is up, east is to the left! Consequently, the position angle is measured counter-clockwise.

All double stars with a combined (integrated) magnitude of 6.5 or brighter are plotted on the charts. The star chart tables contain only a selection of the finest targets. People who are especially interested in observing double stars should consider *The Cambridge Double Star Atlas* (see Sources and references, on page 90).

## Open and globular clusters

Open star clusters appear on the star charts as yellow disks with green outlines. Globular clusters are also shown the same way, but with green crosses through their centers. Open clusters are usually found near the plane of the Milky Way, and so they can be found in or close to lighter areas on the charts, representing the brightest parts of the Milky Way. Open clusters are groups of young stars, often hot and bluish, and their individual stars can be seen easily with a small telescope or sometimes with the naked eye. The last column of this section of the star chart table (N*) gives the approximate number of stars in the cluster.

Globular clusters are quite different. They contain larger numbers of stars and are much more compact. They are found outside the Galactic plane and the stars are older. All clusters down to magnitude 10 are plotted on the star charts. The diameter (Diam) in the star chart table is given in minutes (′) of arc.

In the first column of the star chart tables you will find the designation of the cluster. First the NGC or IC numbers are listed (NGC stands for the *New General Catalogue*; IC for *Index Catalogue* or the *Second Index Catalogue*). NGC numbers are without a prefix (both on the charts and in the tables), and IC numbers have the prefix 'I.'. Alternative names, as well as Messier numbers (M), are in the second column. The same goes for nebulae and galaxies.

## Diffuse and planetary nebulae

Diffuse nebulae are areas of the raw materials, dust

and gas, from which stars are born. These diffuse nebulae are also found along the spiral arms of the Milky Way, and are visible in other nearby galaxies as well. There are three general types of diffuse nebulae. In an *emission nebulae* (E) one or more hot stars causes the nebula to emit light of its own. A *reflection nebula* (R) only reflects light from nearby stars. Reflection nebulae show a blue color, in contrast to the reddish glow of emission nebulae. The third type is the dark nebula. Most dark nebulae are not easily found in amateur telescopes. Only one is shown on the maps: the Coal Sack in the constellation Crux (Southern Cross), a dark patch in the Milky Way visible with the naked eye.

Planetary nebulae have nothing to do with planets. The name arose from their disk-like appearance. They are almost spherical cast-off shells of gas from very hot stars, late in their life-spans. Often the ionized gas has a greenish color. Examples are the Ring Nebula (M57) in Lyra (the Lyre) and the Helix Nebula (NGC 7293) in Aquarius (the Water Carrier).

Both bright diffuse nebulae and planetary nebulae are shown in soft green on the atlas charts. The larger nebulae are drawn to scale, and the smaller ones are shown as small green boxes.

## Galaxies

The red ovals on the charts are the most remote objects: the galaxies. Galaxies are huge systems of stars, clusters and nebulae, like our own Milky Way. There are several types of galaxy: the elliptical (E), spiral (S), barred spiral (SB), and irregular. The E type is subdivided according to shape: from E0 for the almost spherical to E7 for the flattened lens shape. The S and SB types are subdivided according to how tightly the spiral arms are wound (see figure 6).

**Figure 6**

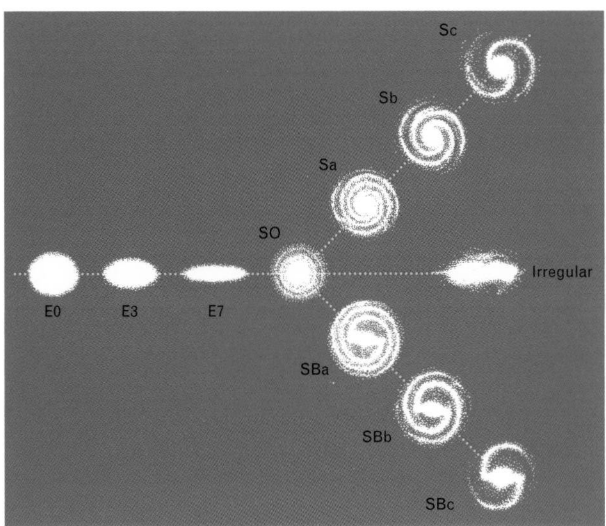


## The Galactic Equator

On the charts the Galactic Equator is drawn as a blue dashed line. It represents the projection of the plane of our galaxy on the stellar sphere. Every ten degrees of galactic longitude is marked along this equator. The center of our Milky Way is at 0° galactic longitude, and can be found on chart 18, in Sagittarius, close to the Ophiuchus boundary.

## The ecliptic

The *ecliptic* is the projection of the Earth's orbit around the Sun, or the yearly path of the Sun along the sky, caused by the orbital movement of the Earth. It is also shown on the charts as a green dashed line. Every ten degrees of longitude is marked. The ecliptic longitude starts at the vernal equinox, at right ascension $0^h$ and declination 0° (charts 8 and 13). The vernal equinox marks the Sun's position when it is exactly in the plane of the Earth's Equator, around March 21 each year.

## The Moon and the planets

The Moon and the planets will not be found on a regular star chart, and the charts in this atlas are no exception. Like the Earth, the other planets of our solar system orbit the Sun, and the Moon orbits the Earth. For that reason they are always changing their position in relation to the background of 'fixed' stars. But since the orbits of the Moon and the planets, including the Earth, lie almost in the same plane, they can always be found close to the ecliptic.

If you want actual information on where to find the planets you can consult one of the many astronomical yearbooks or magazines that are available. Astronomical computer programs can also be very helpful by showing their positions in the sky for any date and time.

## Abbreviations used in the star chart tables

A table, giving information for all of the interesting telescopic objects on the charts, accompanies each star chart. The listing is restricted to the main regions of each chart excluding areas of overlap. These regions are indicated in each table heading and are shown on the Index to the star charts on the facing page, as well as on the Messier charts on pages 22–29. The constellation abbreviations follow the official IAU practice (see the list of constellations in table E).

For the double stars only a selection is shown, limited to double stars that can be separated in small or medium-sized telescopes. As far as possible all plotted variable stars, open and globular clusters, planetary and diffuse nebulae, and galaxies are listed. All positions are for the epoch 2000.0.

In general visual (V) magnitudes are listed, except for the planetary nebulae, where the photographic magnitudes are given. Elsewhere, when a photographic magnitude is used, the figure is followed by 'p'.

When a colon follows an entry in the tables, it means that the given value is uncertain.

### General

**Con** Constellation (the official IAU abbreviations: see table E, on page 35)
**RA** Right ascension
**Dec** Declination
**NGC/IC** Numbers from the *New General Catalogue* (NGC) or *Index Catalogues* (IC) by J.L.E. Dryer. NGC numbers have no prefix, IC numbers have the prefix I.

### Variable stars

Range (max - min), type, period (days), and spectrum are given. In general only variable stars with a range of more than 0.5 magnitudes are plotted and listed.

**Type** Type of variable star
- **Cep** Classical Cepheid
- **CW** Type II Cepheid
- **E** Eclipsing binary
- **EA** Algol type
- **EB** Beta Lyrae type
- **EW** W Ursae Majoris type
- **M** Mira (long-period) type
- **SR** Semi-regular
- **Irr** Irregular
- **RCB** R Coronae Borealis type
- **δ Sct** Delta Scuti type
- **ZA** Z Andromedae type
- **Rn** Recurrent nova
- **N** Nova
- **RV** RV Tauri type
- **SD** S Dor type

### Double stars

In general double stars with a separation less than 1 minute of arc, or with components fainter than magnitude 10.0, are not listed. When a magnitude entry is followed by 'd', it means that this component itself is also a double star.

**PA** Position angle, counter-clockwise from north (0°)
**Sep** Separation, in seconds of arc
The data are for the year 2000

### Open clusters

**Mag** Integrated visual magnitude
**Diam** Diameter, in minutes of arc
**N*** Approximate number of stars

### Globular clusters

**Diam** Diameter, in minutes of arc
**Mag** Integrated visual magnitude

### Bright diffuse nebulae

**Type** Type of nebulae
- **E** Emission nebula
- **R** Reflection nebula

**Diam** Angular dimensions, in minutes of arc
**Mag*** Approximate magnitude of the illuminating star

### Planetary nebulae

**Diam** Diameter, in seconds of arc
Two values separated by a slash (/) refer to a bright core surrounded by a fainter halo
**Mag** Integrated photographic magnitude
**Mag*** Photoelectric magnitude of the central star

### Galaxies

**Mag** Integrated visual magnitude
**Diam** Major and minor diameters, in minutes of arc
**Type** Type of galaxy (see also figure 6 on page 37)
- **E** Elliptical galaxy
  The letter E is usually followed by a number ranging from 0 (almost circular) to 7 (flattened to a lens-shape)
- **Sa** Spiral galaxy with tightly wound arms
- **Sb** Spiral galaxy with less tight arms
- **Sc** Spiral galaxy with loosely wound arms
- **SBa** Barred spiral galaxy with tightly wound spiral arms
- **SBb** Barred spiral galaxy with less tight arms
- **SBc** Barred spiral galaxy with loosely wound arms
- **Irr** Irregular galaxy

# INDEX TO THE STAR CHARTS

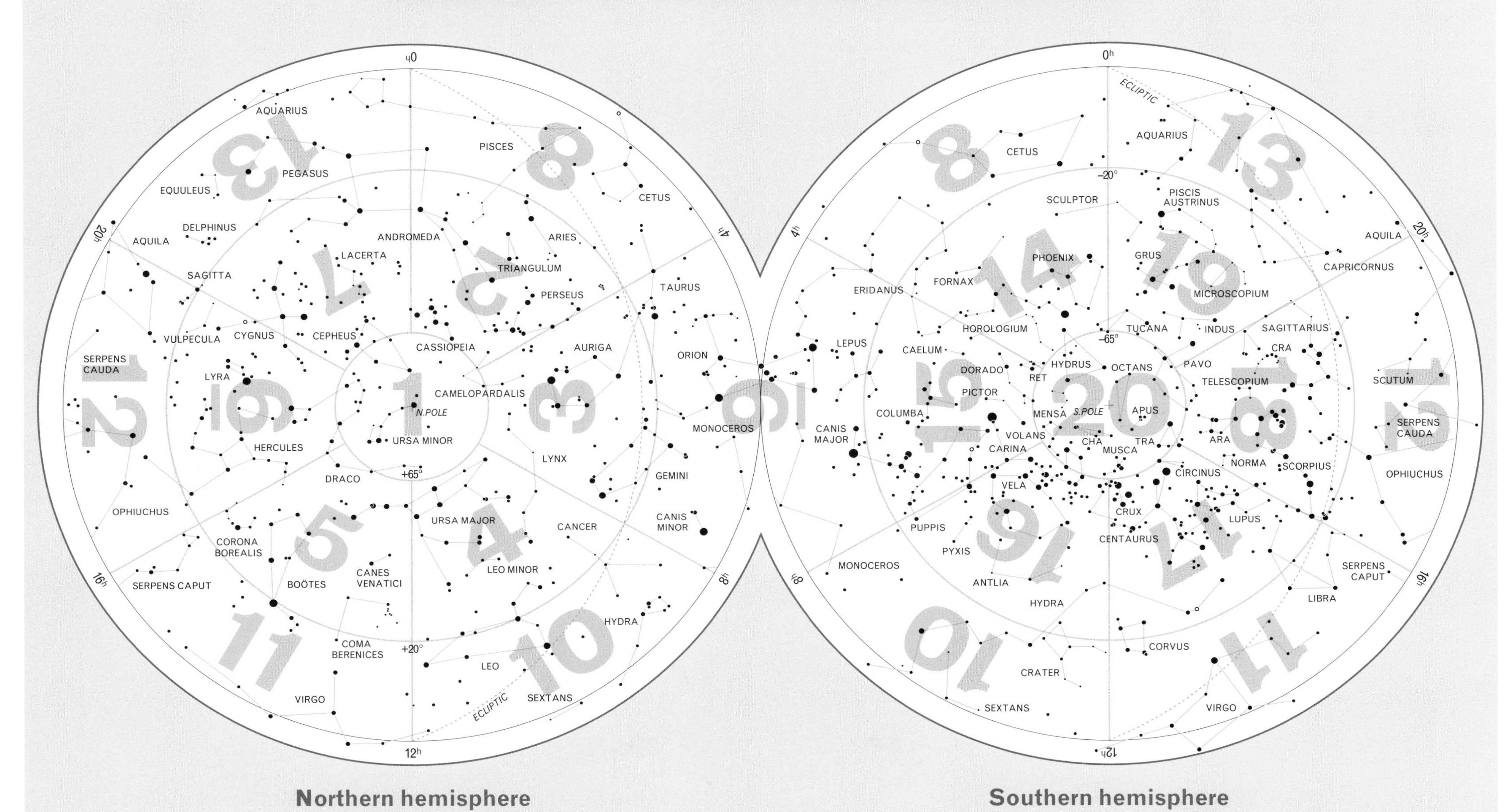


Northern hemisphere

Southern hemisphere

# Chart 1 *North of declination +65°*

## Variable stars

| Star | Con | RA h m | Dec ° ′ | Range | Type | Period (days) | Spectrum |
|---|---|---|---|---|---|---|---|
| YZ | Cas | 00 45.7 | +74 59 | 5.7–6.1 | EA | 4.47 | A + F |
| RZ | Cas | 02 48.9 | +69 38 | 6.2–7.7 | EA | 1.20 | A |
| R | UMa | 10 44.6 | +68 47 | 6.7–13.4 | M | 301.7 | M |
| VY | UMa | 10 45.1 | +67 25 | 5.9–6.5 | Irr | — | M |
| RY | Dra | 12 56.4 | +66 00 | 6.0–8.0 | SR | 172.5: | K |
| RR | UMi | 14 57.6 | +65 56 | 6.1–6.5 p | SR | 40: | M |
| UX | Dra | 19 21.6 | +76 34 | 5.9–7.1 | SR | 168: | C |

## Double stars

| Star | Con | RA h m | Dec ° ′ | PA ° | Sep ″ | Magnitudes | |
|---|---|---|---|---|---|---|---|
| ψ | Cas | 01 25.9 | +68 08 | 025 | 33.4 | 4.7 + 9.6 | |
| 48 | Cas | 02 02.0 | +70 54 | 263 | 0.9 | 4.7 + 6.4 | Binary, 60.4 years |
| ι | Cas | 02 29.1 | +67 24 | 230 | 2.5 | 4.6 + 6.9 | Binary, 840 years |
| | | | | 114 | 7.2 | 8.4 | |
| α | UMi | 02 31.8 | +89 16 | 218 | 18.4 | 2.0 + 9.0 | Polaris |
| ψ | Dra | 17 41.9 | +72 09 | 015 | 30.3 | 4.9 + 6.1 | |
| 40, 41 | Dra | 18 00.2 | +80 00 | 232 | 19.3 | 5.7 + 6.1 | |
| ε | Dra | 19 48.2 | +70 16 | 015 | 3.1 | 3.8 + 7.4 | |
| κ | Cep | 20 08.9 | +77 43 | 122 | 7.4 | 4.4 + 8.4 | |
| β | Cep | 21 28.7 | +70 34 | 249 | 13.3 | 3.2 + 7.9 | Alfirk |
| π | Cep | 23 07.9 | +75 23 | 357 | 1.2 | 4.6 + 6.6 | Binary, 147 years |
| ο | Cep | 23 18.6 | +68 07 | 223 | 2.8 | 4.9 + 7.1 | Binary, 796.2 years |

## Open cluster

| NGC/IC | Other | Con | RA h m | Dec ° ′ | Mag | Diam ′ | N* |
|---|---|---|---|---|---|---|---|
| 188 | | Cep | 00 44.4 | +85 20 | 8.1 | 14 | 120 |

## Bright diffuse nebulae

| NGC/IC | Other | Con | RA h m | Dec ° ′ | Type | Diam ′ | Mag* |
|---|---|---|---|---|---|---|---|
| 7822 | | Cep | 00 03.6 | +68 37 | E | 60 × 30 | |
| — | Ced 214 | Cep | 00 04.7 | +67 10 | E + R | 50 × 40 | 5.7 |
| 7023 | | Cep | 21 01.8 | +68 12 | R | 18 × 18 | 6.8 |

## Planetary nebulae

| NGC/IC | Other | Con | RA h m | Dec ° ′ | Mag p | Diam ″ | Mag* |
|---|---|---|---|---|---|---|---|
| 40 | | Cep | 00 13.0 | +72 32 | 10.7 | 37 | 11.6 |
| I.3568 | | Cam | 12 32.9 | +82 33 | 11.6 | 6 | 12.3 |
| 6543 | | Dra | 17 58.6 | +66 38 | 8.8 | 18/350 | 11.4 |

## Galaxies

| NGC/IC | Other | Con | RA h m | Dec ° ′ | Mag | Size ′ | Type | |
|---|---|---|---|---|---|---|---|---|
| I.342 | | Cam | 03 46.8 | +68 06 | 9.1 | 17.8 × 17.4 | Sc | |
| I.356 | | Cam | 04 07.8 | +69 49 | 11.4 | 5.2 × 4.1 | Sb | |
| 1560 | | Cam | 04 32.8 | +71 53 | 11.5 | 9.8 × 2.0 | Sd | |
| 1961 | | Cam | 05 42.1 | +69 23 | 11.1 | 4.3 × 3.0 | Sb | |
| 2146 | | Cam | 06 18.7 | +78 21 | 10.5 | 6.0 × 3.8 | SBb | |
| 2336 | | Cam | 07 27.1 | +80 11 | 10.5 | 6.9 × 4.0 | Sb | |
| 2366 | | Cam | 07 28.9 | +69 13 | 10.9 | 7.6 × 3.5 | Irr | |
| 2403 | | Cam | 07 36.9 | +65 36 | 8.4 | 17.8 × 11.0 | Sc | |
| — | U4305 | UMa | 08 18.9 | +70 43 | 10.6 | 7.6 × 6.2 | Irr | |
| 2655 | | Cam | 08 55.6 | +78 13 | 10.1 | 5.1 × 4.1 | SBa | |
| 2715 | | Cam | 09 08.1 | +78 05 | 11.4 | 5.0 × 1.9 | Sc | |
| 2787 | | UMa | 09 19.3 | +69 12 | 10.8 | 3.4 × 2.3 | Sa | |
| 2976 | | UMa | 09 47.3 | +67 55 | 10.2 | 4.9 × 2.5 | Sc | |
| 2985 | | UMa | 09 50.4 | +72 17 | 10.5 | 4.3 × 3.4 | Sb | |
| 3031 | M81 | UMa | 09 55.6 | +69 04 | 6.9 | 25.7 × 14.1 | Sb | Bode's Galaxy |
| 3034 | M82 | UMa | 09 55.8 | +69 41 | 8.4 | 11.2 × 4.6 | Pec | Cigar Galaxy |
| 3077 | | UMa | 10 03.3 | +68 44 | 9.9 | 4.6 × 3.6 | E2 | |
| 3147 | | Dra | 10 16.9 | +73 24 | 10.7 | 4.0 × 3.5 | Sb | |
| I.2574 | | UMa | 10 28.4 | +68 25 | 10.6 | 12.3 × 5.9 | S | |
| 3348 | | UMa | 10 47.2 | +72 50 | 11.2 | 2.2 × 2.2 | E1 | |
| 4125 | | Dra | 12 08.1 | +65 11 | 9.8 | 5.1 × 3.2 | E5 | |
| 4236 | | Dra | 12 16.7 | +69 28 | 9.7 | 18.6 × 6.9 | SB | |
| 4589 | | Dra | 12 37.4 | +74 12 | 11.8 | 3.0 × 2.7 | Sa | |
| 4750 | | Dra | 12 50.1 | +72 52 | 11.9 | 2.3 × 2.1 | Sa | |
| 6503 | | Dra | 17 49.4 | +70 09 | 10.2 | 6.2 × 2.3 | Sb | |

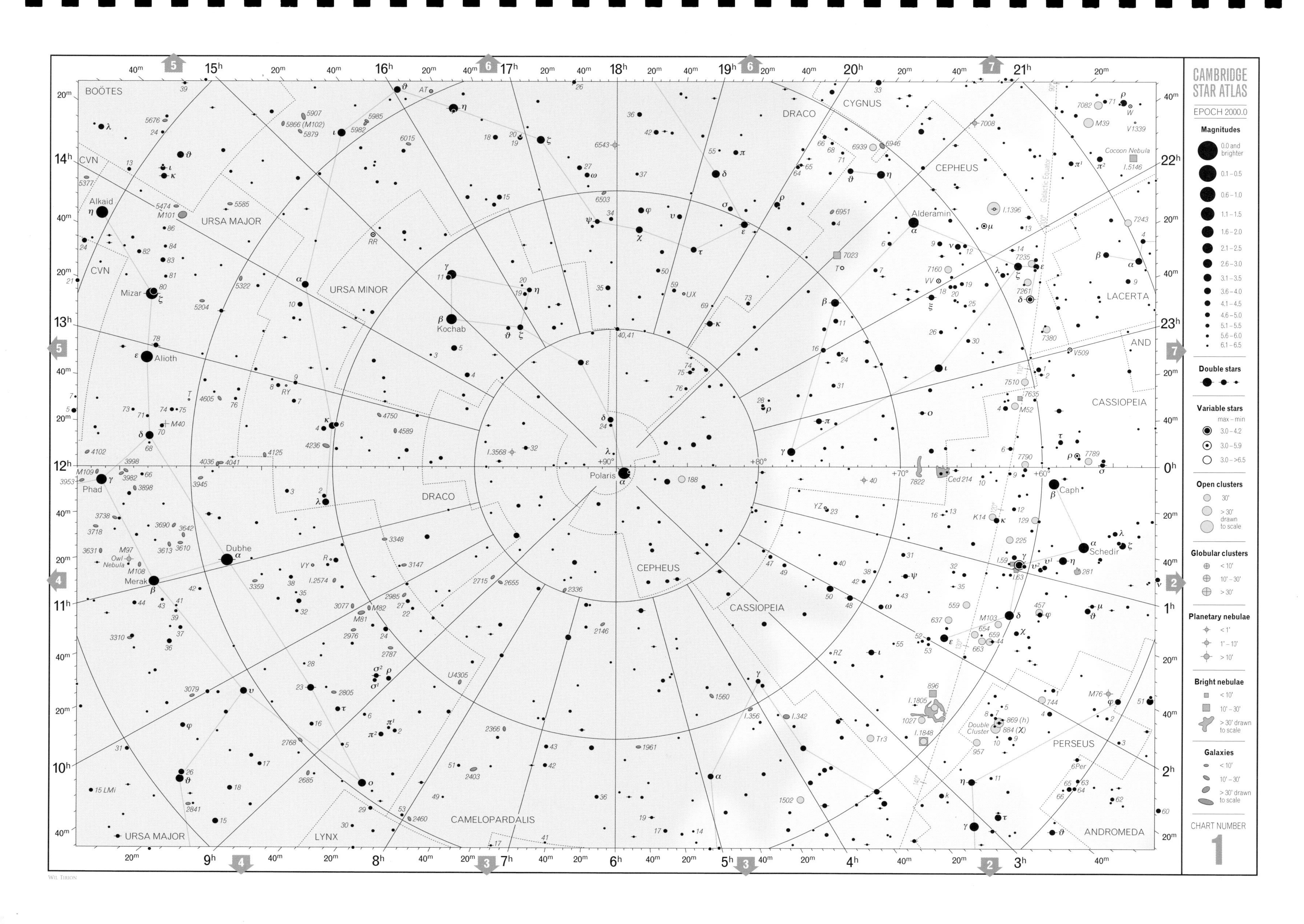

CAMBRIDGE STAR ATLAS
EPOCH 2000.0
Magnitudes
0.0 and brighter
0.1 – 0.5
0.6 – 1.0
1.1 – 1.5
1.6 – 2.0
2.1 – 2.5
2.6 – 3.0
3.1 – 3.5
3.6 – 4.0
4.1 – 4.5
4.6 – 5.0
5.1 – 5.5
5.6 – 6.0
6.1 – 6.5
Double stars
Variable stars
max – min
3.0 – 4.2
3.0 – 5.9
3.0 – >6.5
Open clusters
30'
> 30' drawn to scale
Globular clusters
< 10'
10' – 30'
> 30'
Planetary nebulae
< 1'
1' – 10'
> 10'
Bright nebulae
< 10'
10' – 30'
> 30' drawn to scale
Galaxies
< 10'
10' – 30'
> 30' drawn to scale
CHART NUMBER
1
BOÖTES
CVN
URSA MAJOR
URSA MINOR
DRACO
CYGNUS
CEPHEUS
LACERTA
AND
CASSIOPEIA
PERSEUS
ANDROMEDA
CAMELOPARDALIS
LYNX
Polaris
Kochab
Alderamin
Caph
Schedir
Dubhe
Merak
Phad
Alioth
Mizar
Alkaid
Double Cluster
Cocoon Nebula
Owl Nebula
Galactic Equator
Wil Tirion

## Chart 2 *RA $0^h$ to $4^h$, declination +65° to +20°*

### Variable stars

| Star | Con | RA h m | Dec ° ′ | Range | Type | Period (days) | Spectrum |
|---|---|---|---|---|---|---|---|
| R | **And** | 00 24.0 | +38 35 | 5.8–14.9 | **M** | 409.3 | M |
| γ | **Cas** | 00 56.7 | +60 43 | 1.6–3.0 | **Irr** | — | B |
| R | **Tri** | 02 37.0 | +34 16 | 5.4–12.6 | **M** | 266.5 | M |
| ρ | **Per** | 03 05.2 | +38 50 | 3.3–4.0 | **EA** | 50: | M |
| β | **Per** | 03 08.2 | +40 57 | 2.1–3.4 | **EA** | 2.87 | B + G Algol |
| X | **Per** | 03 55.4 | +31 03 | 6.1–7.0 | **Irr** | — | O |

### Double stars

| Star | Con | RA h m | Dec ° ′ | PA ° | Sep ″ | Magnitudes | |
|---|---|---|---|---|---|---|---|
| π | **And** | 00 36.9 | +33 43 | 173 | 35.9 | 4.4 + 8.6 | |
| η | **Cas** | 00 49.1 | +57 49 | 317 | 12.9 | 3.4 + 7.5 | Binary, 480 years |
| γ | **Cas** | 00 56.7 | +60 43 | 248 | 2.1 | 1.6 v + 11.2 | Variable |
| ψ¹ | **Psc** | 01 05.6 | +21 28 | 159 | 30.0 | 5.6 + 5.8 | |
| 35 | **Cas** | 01 21.1 | +64 40 | 344 | 55.5 | 6.3 + 8.7 | |
| γ | **And** | 02 03.9 | +42 20 | 63 | 9.8 | 2.3 + 5.1 d | Almaak |
| | | | | 103 | 0.4 | 5.5 + 6.3 | Binary, 61.1 years |
| 6 | **Tri** | 02 12.4 | +30 18 | 71 | 3.9 | 5.3 + 6.9 | = ε Tri |
| 30 | **Ari** | 02 37.0 | +24 39 | 274 | 38.6 | 6.6 + 7.4 | |
| 33 | **Ari** | 02 40.7 | +27 04 | 0 | 28.6 | 5.5 + 8.4 | |
| ε | **Ari** | 02 59.2 | +21 20 | 203 | 1.4 | 5.2 + 5.5 | |
| ϑ | **Per** | 02 44.2 | +49 14 | 305 | 20.0 | 4.1 + 9.9 | Binary, 2720 years |
| η | **Per** | 02 50.7 | +55 54 | 300 | 28.3 | 3.8 + 8.5 | |
| | | | | 268 | 66.6 | 9.8 | |
| 40 | **Per** | 03 42.4 | +33 58 | 238 | 20.0 | 5.0 + 9.5 | |
| ο | **Per** | 03 44.3 | +32 17 | 37 | 10 | 3.8 + 8.3 | |
| ζ | **Per** | 03 54.1 | +31 53 | 208 | 12.9 | 2.9 + 9.5 | Atik |
| | | | | 286 | 32.8 | 11.3 | |
| | | | | 195 | 94.2 | 9.5 | |
| | | | | 185 | 120.3 | 10.2 | |
| ε | **Per** | 03 57.9 | +40 01 | 10 | 8.8 | 2.9 + 8.1 | |

### Open clusters

| NGC/IC | Other | Con | RA h m | Dec ° ′ | Mag | Diam ′ | N* | |
|---|---|---|---|---|---|---|---|---|
| 129 | | **Cas** | 00 29.9 | +60 14 | 6.5 | 21 | 35 | |
| — | K14 | **Cas** | 00 31.9 | +63 10 | 8.5 | 7 | 20 | |
| 225 | | **Cas** | 00 43.4 | +61 47 | 7.0 | 12 | 15 | |
| 457 | | **Cas** | 01 19.1 | +58 20 | 6.4 | 13 | 80 | |
| 559 | | **Cas** | 01 29.5 | +63 18 | 9.5 | 4.4 | 60 | |
| 581 | M103 | **Cas** | 01 33.2 | +60 42 | 7.4 | 6 | 25 | |
| 637 | | **Cas** | 01 42.9 | +64 00 | 8.2 | 3.5 | 20 | |
| 654 | | **Cas** | 01 44.1 | +61 53 | 6.5 | 5 | 60 | |
| 659 | | **Cas** | 01 44.2 | +60 42 | 7.9 | 5 | 40 | |
| 663 | | **Cas** | 01 46.0 | +61 15 | 7.1 | 16 | 80 | |
| 752 | | **And** | 01 57.8 | +37 41 | 5.7 | 50 | 60 | |
| 744 | | **Per** | 01 58.4 | +55 29 | 7.9 | 11 | 20 | |
| 869 | h | **Per** | 02 19.0 | +57 09 | 4.3 p | 30 | 200 | Double Cluster |
| 884 | χ | **Per** | 02 22.4 | +57 07 | 4.4 p | 30 | 150 | Double Cluster |

### Open clusters *(continued)*

| NGC/IC | Other | Con | RA h m | Dec ° ′ | Mag | Diam ′ | N* | |
|---|---|---|---|---|---|---|---|---|
| I.1805 | | **Cas** | 02 32.7 | +61 27 | 6.5 | 22 | 40 | In nebula |
| 957 | | **Per** | 02 33.6 | +57 32 | 7.6 | 11 | 30 | |
| 1039 | M34 | **Per** | 02 42.0 | +42 47 | 5.2 | 35 | 60 | |
| 1027 | | **Cas** | 02 42.7 | +61 33 | 6.7 | 20 | 40 | |
| I.1848 | | **Cas** | 02 51.2 | +60 26 | 6.5 | 12 | 10 | |
| — | Tr 3 | **Cas** | 03 11.8 | +63 15 | 7.0 p | 23 | 30 | |
| 1245 | | **Per** | 03 14.7 | +47 15 | 8.4 | 10 | 200 | |
| 1342 | | **Per** | 03 31.6 | +37 20 | 6.7 | 14 | 40 | |
| — | M45 | **Tau** | 03 47.0 | +24 07 | 1.2 | 110 | 100 | Pleiades |
| 1444 | | **Per** | 03 49.4 | +52 40 | 6.6 | 4.0 | | |

### Bright diffuse nebulae

| NGC/IC | Other | Con | RA h m | Dec ° ′ | Type | Diam ′ | Mag* | |
|---|---|---|---|---|---|---|---|---|
| 281 | | **Cas** | 00 52.8 | +56 36 | **E** | 35 × 30 | 7.8 | |
| I.59 | | **Cas** | 00 56.7 | +61 04 | **E + R** | 10 × 5 | 2.5 | γ Cas |
| I.63 | | **Cas** | 00 59.5 | +60 49 | **E + R** | 10 × 3 | 2.5 | γ Cas |
| 896 | | **Cas** | 02 24.8 | +61 54 | **E** | 27 × 13 | 10.5 | |
| I.1805 | | **Cas** | 02 33.4 | +61 26 | **E** | 60 × 60 | | |
| I.1848 | | **Cas** | 02 51.3 | +60 25 | **E** | 60 × 30 | | |

### Planetary nebula

| NGC/IC | Other | Con | RA h m | Dec ° ′ | Mag p | Diam ″ | Mag* | |
|---|---|---|---|---|---|---|---|---|
| 650, 651 | M76 | **Per** | 01 42.4 | +51 34 | 12.2 | 65/290 | 17.0: | Little Dumbbell |

### Galaxies

| NGC/IC | Other | Con | RA h m | Dec ° ′ | Mag | Size ′ | Type | |
|---|---|---|---|---|---|---|---|---|
| 23 | | **Peg** | 00 09.9 | +25 55 | 11.9 | 2.3 × 1.6 | **Sc** | |
| 147 | | **Cas** | 00 33.2 | +48 30 | 9.3 | 12.9 × 8.1 | **dE4** | |
| 185 | | **Cas** | 00 39.0 | +48 20 | 9.2 | 11.5 × 9.8 | **dE0** | |
| 205 | M110 | **And** | 00 40.4 | +41 41 | 8.0 | 17.4 × 9.8 | **E6** | Companion to M31 |
| 221 | M32 | **And** | 00 42.7 | +40 52 | 8.2 | 7.6 × 5.8 | **E2** | Companion to M31 |
| 224 | M31 | **And** | 00 42.7 | +41 16 | 3.5 | 178 × 63 | **Sb** | Andromeda Galaxy |
| 278 | | **Cas** | 00 52.1 | +47 33 | 10.9 | 2.2 × 2.1 | **E0** | |
| 404 | | **And** | 01 09.4 | +35 43 | 10.1 | 4.4 × 4.2 | **E0** | Appr. 6.5′ NW of β And |
| 598 | M33 | **Tri** | 01 33.9 | +30 39 | 5.7 | 62 × 39 | **Sc** | Triangulum Galaxy |
| I.1727 | | **Tri** | 01 47.5 | +27 20 | 11.6 | 6.2 × 2.9 | **SB** | |
| 672 | | **Tri** | 01 47.9 | +27 26 | 10.8 | 6.6 × 2.7 | **SBc** | |
| 784 | | **Tri** | 02 01.3 | +28 50 | 11.8 | 6.2 × 1.7 | **SB** | |
| 891 | | **And** | 02 22.6 | +42 21 | 10.0 | 13.5 × 2.8 | **Sb** | |
| 925 | | **Tri** | 02 27.3 | +33 35 | 10.0 | 9.8 × 6.0 | **SBc** | |
| I.239 | | **And** | 02 36.5 | +38 58 | 11.2 | 4.6 × 4.3 | **SBc** | |
| 1003 | | **Per** | 02 39.3 | +40 52 | 11.5 | 5.4 × 2.1 | **Sc** | |
| 1023 | | **Per** | 02 40.4 | +39 04 | 9.5 | 8.7 × 3.3 | **E7** | |

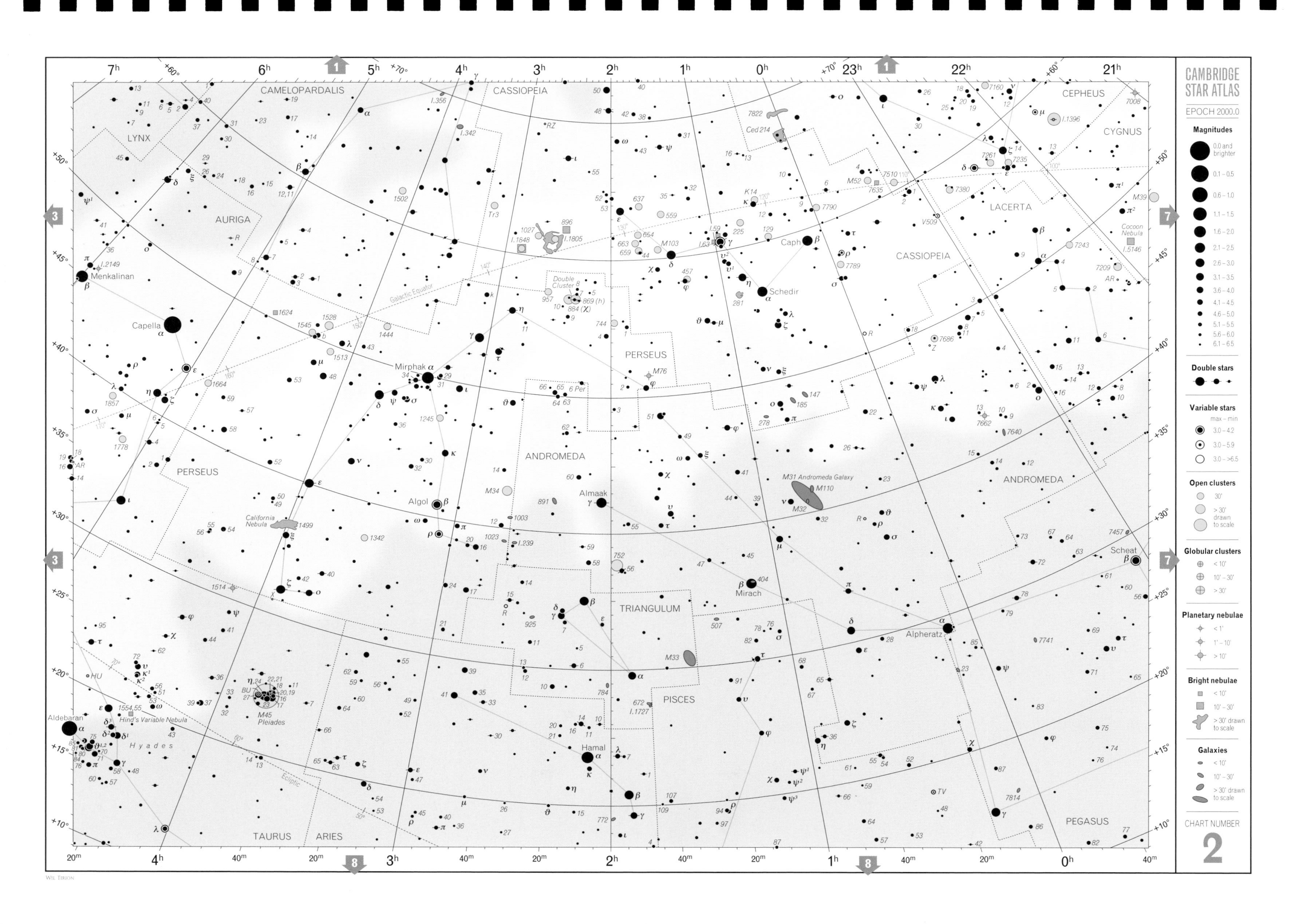

CAMBRIDGE STAR ATLAS
EPOCH 2000.0
Magnitudes
0.0 and brighter
0.1 – 0.5
0.6 – 1.0
1.1 – 1.5
1.6 – 2.0
2.1 – 2.5
2.6 – 3.0
3.1 – 3.5
3.6 – 4.0
4.1 – 4.5
4.6 – 5.0
5.1 – 5.5
5.6 – 6.0
6.1 – 6.5
Double stars
Variable stars
max – min
3.0 – 4.2
3.0 – 5.9
3.0 – >6.5
Open clusters
30'
> 30' drawn to scale
Globular clusters
< 10'
10' – 30'
> 30'
Planetary nebulae
< 1'
1' – 10'
> 10'
Bright nebulae
< 10'
10' – 30'
> 30' drawn to scale
Galaxies
< 10'
10' – 30'
> 30' drawn to scale
CHART NUMBER
2
CAMELOPARDALIS
LYNX
AURIGA
PERSEUS
CASSIOPEIA
CEPHEUS
CYGNUS
LACERTA
ANDROMEDA
TRIANGULUM
PISCES
PEGASUS
TAURUS
ARIES
Capella
Menkalinan
Mirphak
Algol
Almaak
Mirach
Alpheratz
Scheat
Hamal
Aldebaran
Caph
Schedir
Double Cluster
869 (h)
884 (χ)
California Nebula
Cocoon Nebula
M31 Andromeda Galaxy
M110
M32
M33
M34
M45 Pleiades
M76
M103
M52
M39
Hyades
Hind's Variable Nebula
Galactic Equator
Ecliptic
WIL TIRION

# Chart 3 *RA 4h to 8h, declination +65° to +20°*

## Variable stars

| Star | Con | RA h m | Dec ° ′ | Range | Type | Period (days) | Spectrum | |
|---|---|---|---|---|---|---|---|---|
| HU | Tau | 04 38.3 | +20 41 | 5.9–6.7 | EA | 2.06 | A | |
| ε | Aur | 05 02.0 | +43 49 | 2.9–3.8 | EA | 9892 | A – F | |
| R | Aur | 05 17.3 | +53 35 | 6.7–13.7 | M | 457.5 | M | |
| AR | Aur | 05 18.3 | +33 46 | 6.2–6.8 | EA | 4.13 | A + B | |
| U | Ori | 05 55.8 | +20 10 | 4.8–12.6 | M | 372.4 | M | |
| TV | Gem | 06 11.8 | +21 52 | 8.7–9.5 p | SR | 182 | M | |
| BU | Gem | 06 12.3 | +22 54 | 5.7–7.5 | Irr | — | M | |
| η | Gem | 06 14.9 | +22 30 | 3.2–3.9 | SR | 232.9 | M | Propus |
| RT | Aur | 06 28.6 | +30 30 | 5.0–5.8 | Cep | 3.72 | F – G | |
| WW | Aur | 06 32.5 | +32 27 | 5.8–6.5 | EA | 2.53 | A | |
| UU | Aur | 06 36.5 | +38 27 | 7.8–10.0 | SR | 234 | C | |
| ζ | Gem | 07 04.1 | +20 34 | 3.7–4.2 | Cep | 10.15 | F – G | Mekbuda |
| R | Gem | 07 07.4 | +22 42 | 6.0–14.0 | M | 369.8 | S | |

## Double stars

| Star | Con | RA h m | Dec ° ′ | PA ° | Sep ″ | Magnitudes | |
|---|---|---|---|---|---|---|---|
| φ | Tau | 04 20.4 | +27 21 | 250 | 52.1 | 5.0 + 8.0 | |
| χ | Tau | 04 22.6 | +25 38 | 24 | 19.4 | 5.5 + 7.6 | |
| 56 | Per | 04 24.6 | +33 58 | 22 | 4.2 | 5.9 + 8.7 | |
| $\kappa^1$, $\kappa^2$ | Tau | 04 25.4 | +22 18 | 173 | 339 | 4.2 + 5.3 | |
| 4 | Aur | 04 59.3 | +37 53 | 359 | 5.4 | 5.0 + 8.0 | |
| 14 | Aur | 05 15.4 | +32 41 | 352 | 11.1 | 5.1 + 11.1 | |
| | | | | 226 | 14.6 | 7.4 | |
| | | | | 321 | 184.0 | 10.4 | |
| R | Aur | 05 17.3 | +53 35 | 339 | 47.5 | 6.9 v + 8.6 | Variable |
| 118 | Tau | 05 29.3 | +25 09 | 204 | 4.8 | 5.8 + 6.6 | |
| 26 | Aur | 05 38.6 | +30 30 | 3 | 0.2 | 6.0 + 6.3 | Binary, 53.2 years |
| | | | | 267 | 12.4 | 8.0 | |
| ν | Aur | 05 51.5 | +39 09 | 206 | 54.6 | 4.0 + 9.3 | |
| δ | Aur | 05 59.5 | +54 17 | 271 | 115.4 | 3.7 + 9.5 | |
| | | | | 67 | 197.1 | 9.5 | |
| ϑ | Aur | 05 59.7 | +37 13 | 313 | 3.6 | 2.6 + 7.1 | |
| | | | | 297 | 50.0 | 10.6 | |
| η | Gem | 06 14.9 | +22 30 | 257 | 1.6 | 3.3 v + 8.8 | Propus; var.; bin., 473.7 y. |
| μ | Gem | 06 22.9 | +22 31 | 141 | 121.7 | 3.2 + 9.4 | |
| ν | Gem | 06 29.0 | +20 13 | 329 | 112.5 | 4.2 + 8.7 | |
| ε | Gem | 06 43.9 | +25 08 | 94 | 110.3 | 3.0 + 9.0 | Mebsuta |
| 12 | Lyn | 06 46.2 | +59 27 | 70 | 1.7 | 5.4 + 6.0 | Binary, 699 years |
| | | | | 308 | 8.7 | 7.3 | |
| ζ | Gem | 07 04.1 | +20 34 | 84 | 87.0 | 3.7 v + 10.5 | Variable |
| | | | | 350 | 96.5 | 8.0 | |
| δ | Gem | 07 20.1 | +21 59 | 226 | 5.8 | 3.5 + 8.2 | Wasat; binary, 1,200 years |
| 19 | Lyn | 07 22.9 | +55 17 | 315 | 14.8 | 5.6 + 6.5 | |
| | | | | 3 | 214.9 | 8.9 | |
| α | Gem | 07 34.6 | +31 53 | 68 | 4.0 | 1.9 + 2.9 | Castor; binary, 420 years |
| | | | | 163 | 72.5 | 8.8 | |
| 24 | Lyn | 07 43.0 | +58 43 | 320 | 54.7 | 5.0 + 9.5 | |

## Open clusters

| NGC/IC | Other | Con | RA h m | Dec ° ′ | Mag | Diam ′ | N* | |
|---|---|---|---|---|---|---|---|---|
| 1502 | | Cam | 04 07.7 | +62 20 | 5.7 | 8 | 45 | |
| 1513 | | Per | 04 10.0 | +49 31 | 8.4 | 9 | 50 | |
| 1528 | | Per | 04 15.4 | +51 14 | 6.4 | 24 | 40 | |
| 1545 | | Per | 04 20.9 | +50 15 | 6.2 | 18 | 20 | |
| 1664 | | Aur | 04 51.1 | +43 42 | 7.6 | 18 | | |
| 1746 | | Tau | 05 03.6 | +23 49 | 6.1 | 42 | 20 | |
| 1778 | | Aur | 05 08.1 | +37 03 | 7.7 | 7 | 25 | |
| 1857 | | Aur | 05 20.2 | +39 21 | 7.0 | 6 | 40 | |
| 1893 | | Aur | 05 22.7 | +33 24 | 7.5 | 11 | 60 | |
| 1907 | | Aur | 05 28.0 | +35 19 | 8.2 | 7 | 30 | |
| 1912 | M38 | Aur | 05 28.7 | +35 50 | 6.4 | 21 | 100 | |
| 1960 | M36 | Aur | 05 36.1 | +34 08 | 6.0 | 12 | 60 | |
| 2099 | M37 | Aur | 05 52.4 | +32 33 | 5.6 | 24 | 150 | |
| 2129 | | Gem | 06 01.0 | +23 18 | 6.7 | 7 | 40 | |
| I.2157 | | Gem | 06 05.0 | +24 00 | 8.4 | 7 | 20 | |
| 2158 | | Gem | 06 07.5 | +24 06 | 8.6 | 5 | | |
| 2168 | M35 | Gem | 06 08.9 | +24 20 | 5.1 | 28 | 200 | |
| 2175 | | Ori | 06 09.8 | +20 19 | 6.8 | 18 | 60 | In nebula 2174 |
| 2281 | | Aur | 06 49.3 | +41 04 | 5.4 | 15 | 30 | |
| 2331 | | Gem | 07 07.2 | +27 21 | 8.5 | 18 | 30 | |
| 2420 | | Gem | 07 38.5 | +21 34 | 8.3 | 10 | 100 | |

## Globular clusters

| NGC/IC | Other | Con | RA h m | Dec ° ′ | Mag | Diam ′ |
|---|---|---|---|---|---|---|
| 2419 | | Lyn | 07 38.1 | +38 53 | 10.4 | 4.1 |

## Bright diffuse nebulae

| NGC/IC | Other | Con | RA h m | Dec ° ′ | Type | Diam ′ | Mag* | |
|---|---|---|---|---|---|---|---|---|
| 1499 | | Per | 04 00.7 | +36 37 | E | 145 × 40 | 4.0 | California Nebula |
| 1624 | | Per | 04 40.5 | +50 27 | E | 5 × 5 | | |
| 1931 | | Aur | 05 31.4 | +34 15 | E + R | 3 × 3 | | |
| 1952 | M1 | Tau | 05 34.5 | +22 01 | E | 6 × 4 | 16 | Crab Nebula; SNR |
| 2174 | | Gem | 06 09.7 | +20 30 | E | 40 × 30 | 7.6 | Contains cluster 2175 |
| I.443 | | Gem | 06 16.9 | +22 47 | E | 50 × 40 | 8.8 | Supernova remnant |

## Planetary nebulae

| NGC/IC | Other | Con | RA h m | Dec ° ′ | Mag p | Diam ″ | Mag* | |
|---|---|---|---|---|---|---|---|---|
| 1514 | | Tau | 04 09.2 | +30 47 | 10.0 | 114 | 9.4 | |
| 2392 | | Gem | 07 29.2 | +20 55 | 9.9 | 13/44 | 10.5 | Eskimo Nebula |

## Galaxies

| NGC/IC | Other | Con | RA h m | Dec ° ′ | Mag | Size ′ | Type |
|---|---|---|---|---|---|---|---|
| 2460 | | Cam | 07 56.9 | +60 21 | 11.7 | 2.9 × 2.2 | Sb |

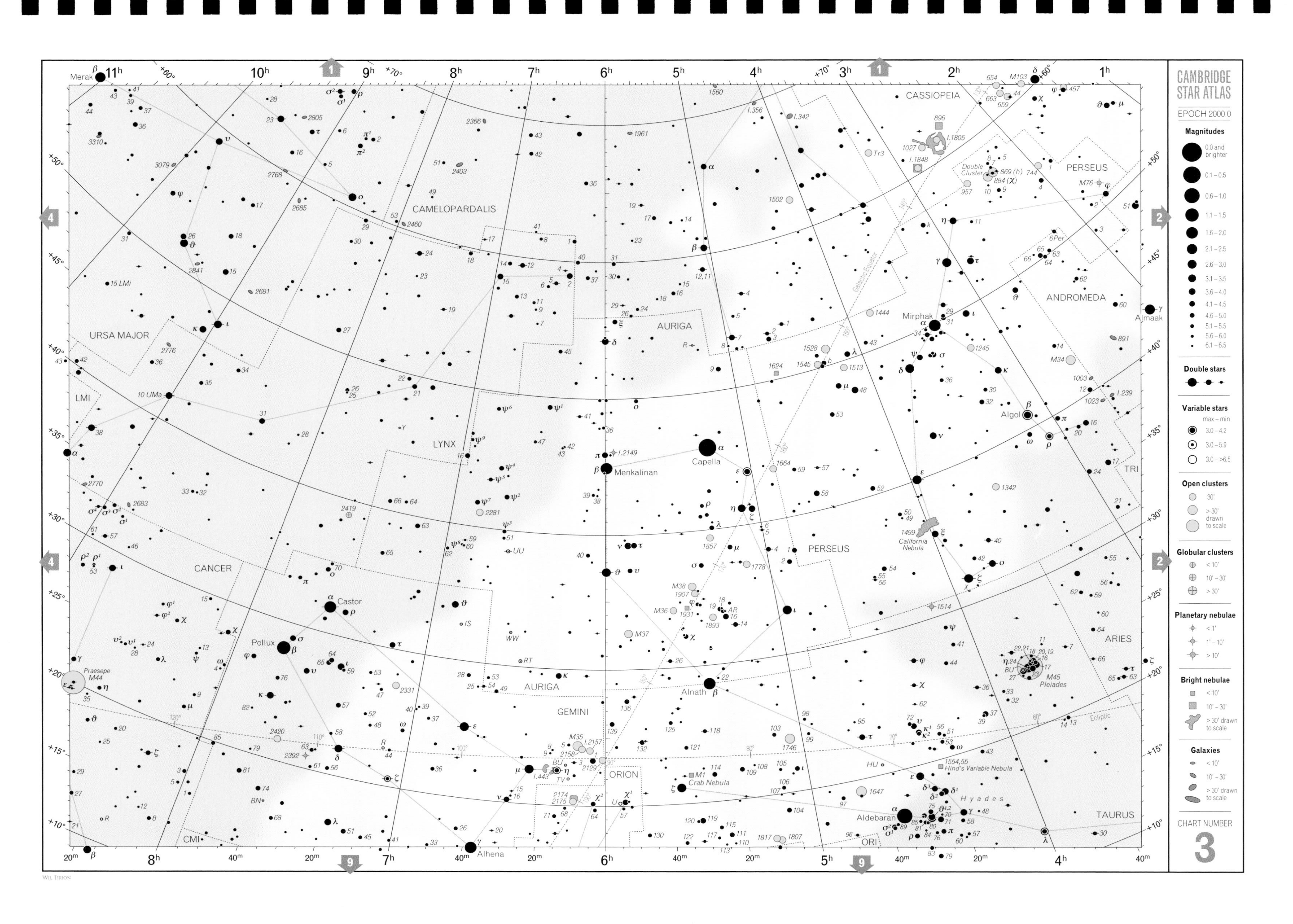
CAMBRIDGE STAR ATLAS
EPOCH 2000.0
Magnitudes
0.0 and brighter
0.1 – 0.5
0.6 – 1.0
1.1 – 1.5
1.6 – 2.0
2.1 – 2.5
2.6 – 3.0
3.1 – 3.5
3.6 – 4.0
4.1 – 4.5
4.6 – 5.0
5.1 – 5.5
5.6 – 6.0
6.1 – 6.5
Double stars
Variable stars
max – min
3.0 – 4.2
3.0 – 5.9
3.0 – >6.5
Open clusters
30'
> 30' drawn to scale
Globular clusters
< 10'
10' – 30'
> 30'
Planetary nebulae
< 1'
1' – 10'
> 10'
Bright nebulae
< 10'
10' – 30'
> 30' drawn to scale
Galaxies
< 10'
10' – 30'
> 30' drawn to scale
CHART NUMBER
3
CASSIOPEIA
PERSEUS
ANDROMEDA
TRI
ARIES
TAURUS
AURIGA
CAMELOPARDALIS
LYNX
GEMINI
ORION
ORI
CANCER
URSA MAJOR
LMI
CMI
Merak
Capella
Menkalinan
Mirphak
Algol
Almaak
Aldebaran
Alnath
Castor
Pollux
Alhena
M45 Pleiades
Hyades
Hind's Variable Nebula
California Nebula
Crab Nebula
Praesepe M44
Double Cluster
Galactic Equator
Ecliptic
Wil Tirion

## Variable stars

| Star | Con | RA h m | Dec ° ′ | Range | Type | Period (days) | Spectrum |
|---|---|---|---|---|---|---|---|
| RS | Cnc | 09 10.6 | +30 58 | 5.1–7.0 | SR | 120: | M |
| R | LMi | 09 45.6 | +34 31 | 6.3–13.2 | M | 371.9 | M |
| ST | UMa | 11 27.8 | +45 11 | 7.7–9.5 | SR | 81 | M |

## Double stars

| Star | Con | RA h m | Dec ° ′ | PA ° | Sep ″ | Magnitudes | |
|---|---|---|---|---|---|---|---|
| ι | Cnc | 08 46.7 | +28 46 | 307 | 30.5 | 4.2 + 6.6 | |
| ι | UMa | 08 59.2 | +48 02 | 177 | 2.0 | 3.1 + 10.2 | Talitha; binary, 817.9 years |
| 38 | Lyn | 09 18.8 | +36 48 | 229 | 2.7 | 3.9 + 6.6 | |
| 23 | UMa | 09 31.5 | +63 04 | 270 | 22.7 | 3.7 + 8.9 | |
| 54 | Leo | 10 55.6 | +24 45 | 110 | 6.5 | 4.5 + 6.3 | |
| α | UMa | 11 03.7 | +61 45 | 223 | 0.6 | 1.9 + 4.8 | Dubhe; binary, 44.7 years |
| ξ | UMa | 11 18.2 | +31 32 | 273 | 1.8 | 4.3 + 4.8 | Binary, 59.8 years |
| ν | UMa | 11 18.5 | +33 06 | 147 | 7.2 | 3.5 + 9.9 | |
| 57 | UMa | 11 29.1 | +39 20 | 359 | 5.4 | 5.3 + 8.3 | |

## Planetary nebula

| NGC/IC | Other | Con | RA h m | Dec ° ′ | Mag p | Diam ″ | Mag* | |
|---|---|---|---|---|---|---|---|---|
| 3587 | M97 | UMa | 11 14.8 | +55 01 | 12.0 | 194 | 15.9 | Owl Nebula |

## Galaxies

| NGC/IC | Other | Con | RA h m | Dec ° ′ | Mag | Size ′ | Type | |
|---|---|---|---|---|---|---|---|---|
| 2683 | | Lyn | 08 52.7 | +33 25 | 9.7 | 9.3 × 2.5 | Sb | |
| 2681 | | UMa | 08 53.5 | +51 19 | 10.3 | 3.8 × 3.5 | Sa | |
| 2685 | | UMa | 08 55.6 | +58 44 | 11.1 | 5.2 × 3.0 | Sb | |
| 2770 | | Lyn | 09 09.6 | +33 07 | 12.1 | 3.7 × 1.3 | Sc | |
| 2768 | | UMa | 09 11.6 | +60 02 | 10.0 | 6.3 × 2.8 | E5 | |
| 2776 | | Lyn | 09 12.2 | +44 57 | 11.6 | 2.9 × 2.7 | Sc | |
| 2805 | | UMa | 09 20.3 | +64 06 | 11.3 | 6.3 × 5.0 | SBd | |
| 2841 | | UMa | 09 22.0 | +50 58 | 9.3 | 8.1 × 3.8 | Sb | |
| 2903 | | Leo | 09 32.2 | +21 30 | 8.9 | 12.6 × 6.6 | Sb | |
| 3079 | | UMa | 10 02.0 | +55 41 | 10.7 | 7.6 × 1.7 | Sb | |
| 3190 | | Leo | 10 18.1 | +21 50 | 11.0 | 4.6 × 1.8 | Sb | |
| 3184 | | UMa | 10 18.3 | +41 25 | 9.8 | 6.9 × 6.8 | Sc | |
| 3193 | | Leo | 10 18.4 | +21 54 | 10.9 | 2.8 × 2.6 | E0 | |
| 3198 | | UMa | 10 19.9 | +45 33 | 10.4 | 8.3 × 3.7 | Sc | |
| 3245 | | LMi | 10 27.3 | +28 30 | 10.8 | 3.2 × 1.9 | E5 | |
| 3294 | | LMi | 10 36.3 | +37 20 | 11.7 | 3.3 × 1.8 | Sc | |
| 3310 | | UMa | 10 38.7 | +53 30 | 10.9 | 3.6 × 3.0 | SBb | |
| 3319 | | UMa | 10 39.2 | +41 41 | 11.3 | 6.8 × 3.9 | SBc | |
| 3344 | | LMi | 10 43.5 | +24 55 | 10.0 | 6.9 × 6.5 | Sc | |
| 3359 | | UMa | 10 46.6 | +63 13 | 10.5 | 6.8 × 4.3 | SBc | |
| 3430 | | LMi | 10 52.2 | +32 57 | 11.5 | 3.9 × 2.3 | Sc | |
| 3432 | | LMi | 10 52.5 | +36 37 | 11.3 | 6.2 × 1.5 | SB | |
| 3486 | | LMi | 11 00.4 | +28 58 | 10.3 | 6.9 × 5.4 | Sc | |
| 3504 | | LMi | 11 03.2 | +27 58 | 11.1 | 2.7 × 2.2 | Sb | |
| 3556 | M108 | UMa | 11 11.5 | +55 40 | 10.1 | 8.3 × 2.5 | Sc | |
| — | U6253 | Leo | 11 13.5 | +22 10 | 11.5 | 14.5 × 12.9 | dE0 | Leo II |
| 3583 | | UMa | 11 14.2 | +48 19 | 11.7 | 2.8 × 2.0 | Sc | |
| 3610 | | UMa | 11 18.4 | +58 47 | 10.8 | 3.2 × 2.5 | E2 | |
| 3613 | | UMa | 11 18.6 | +58 00 | 11.6 | 3.6 × 2.0 | E5 | |
| 3631 | | UMa | 11 21.0 | +53 10 | 10.4 | 4.6 × 4.1 | Sc | |
| 3646 | | Leo | 11 21.7 | +20 10 | 11.2 | 3.9 × 2.6 | Sc | |
| 3642 | | UMa | 11 22.3 | +59 05 | 11.1 | 5.8 × 4.9 | Sc | |
| 3665 | | UMa | 11 24.7 | +38 46 | 10.8 | 3.2 × 2.6 | E2 | |
| 3675 | | UMa | 11 26.1 | +43 35 | 10.9 | 5.9 × 3.2 | Sb | |
| 3690 | | UMa | 11 28.5 | +58 33 | 12.0 | 2.4 × 1.9 | S | |
| 3718 | | UMa | 11 32.6 | +53 04 | 10.5 | 8.7 × 4.5 | SBa | |
| 3726 | | UMa | 11 33.3 | +47 02 | 10.4 | 6.0 × 4.5 | Sc | |
| 3738 | | UMa | 11 35.8 | +54 31 | 11.7 | 2.6 × 2.0 | Pec | |
| 3769 | | UMa | 11 37.7 | +47 54 | 11.8 | 3.2 × 1.1 | Sb | |
| 3877 | | UMa | 11 46.1 | +47 30 | 11.6 | 5.4 × 1.5 | Sb | |
| 3893 | | UMa | 11 48.6 | +48 43 | 11.1 | 4.4 × 2.8 | Sc | |
| 3898 | | UMa | 11 49.2 | +56 05 | 10.8 | 4.4 × 2.6 | Sb | |
| 3941 | | UMa | 11 52.9 | +36 59 | 11.4 | 3.8 × 2.5 | E3 | |
| 3945 | | UMa | 11 53.2 | +60 41 | 10.6 | 5.5 × 3.6 | SBa | |
| 3949 | | UMa | 11 53.7 | +47 52 | 11.0 | 3.0 × 1.8 | Sb | |
| 3953 | | UMa | 11 53.8 | +52 20 | 10.1 | 6.6 × 3.6 | Sb | |
| 3982 | | UMa | 11 56.5 | +55 08 | 11.7 | 2.5 × 2.2 | Sb | |
| 3992 | M109 | UMa | 11 57.6 | +53 23 | 9.8 | 7.6 × 4.9 | SBb | |
| 3998 | | UMa | 11 57.9 | +55 27 | 10.6 | 3.1 × 2.5 | E2 | |
| I.750 | | UMa | 11 58.9 | +42 43 | 11.8 | 2.9 × 1.4 | Sb | |
| 4026 | | UMa | 11 59.4 | +50 58 | 11.7 | 5.1 × 1.4 | S0 | |

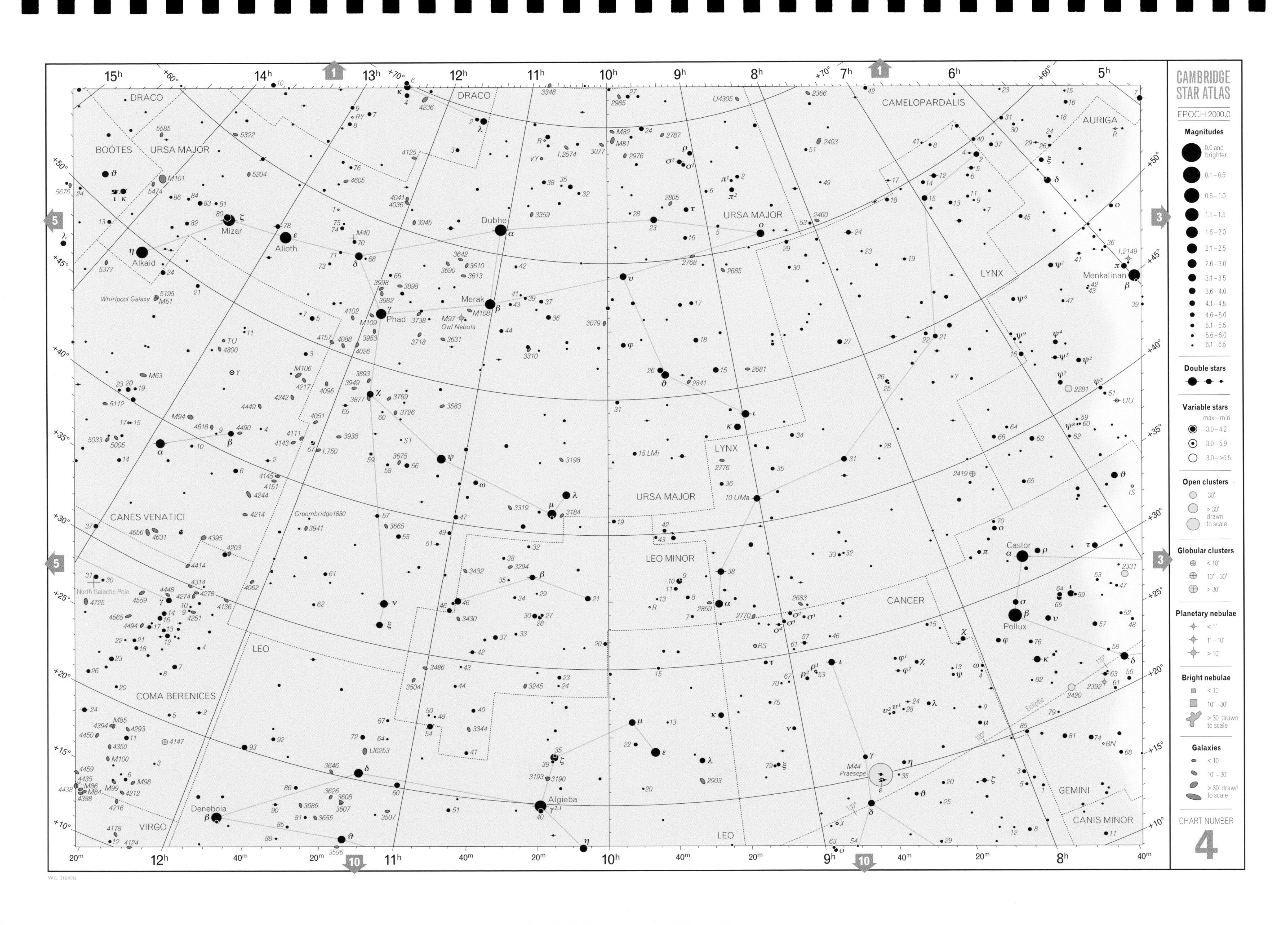
CAMBRIDGE STAR ATLAS
EPOCH 2000.0
Magnitudes
0.0 and brighter
0.1 – 0.5
0.6 – 1.0
1.1 – 1.5
1.6 – 2.0
2.1 – 2.5
2.6 – 3.0
3.1 – 3.5
3.6 – 4.0
4.1 – 4.5
4.6 – 5.0
5.1 – 5.5
5.6 – 6.0
6.1 – 6.5
Double stars
Variable stars
max – min
3.0 – 4.2
3.0 – 5.9
3.0 – >6.5
Open clusters
30'
> 30'
> 30' drawn to scale
Globular clusters
< 10'
10' – 30'
> 30'
Planetary nebulae
< 1'
1' – 10'
> 10'
Bright nebulae
< 10'
10' – 30'
> 30' drawn to scale
Galaxies
< 10'
10' – 30'
> 30' drawn to scale
CHART NUMBER
4
DRACO
BOÖTES
URSA MAJOR
CAMELOPARDALIS
AURIGA
LYNX
CANES VENATICI
LEO MINOR
CANCER
GEMINI
CANIS MINOR
COMA BERENICES
LEO
VIRGO
Mizar
Alioth
Alkaid
Dubhe
Merak
Phad
Menkalinan
Castor
Pollux
Algieba
Denebola
Whirlpool Galaxy
Owl Nebula
M44 Praesepe
North Galactic Pole
Groombridge1830
Ecliptic
WIL TIRION

## Variable stars

| Star | Con | RA h m | Dec ° ′ | Range | Type | Period (days) | Spectrum |
|---|---|---|---|---|---|---|---|
| T | UMa | 12 36.4 | +59 29 | 6.6–13.4 | M | 256.5 | M |
| Y | CVn | 12 45.1 | +45 26 | 5.2–6.6 | SR | 257 | K |
| TU | CVn | 12 54.9 | +47 12 | 5.6–6.6 | SR | 50 | M |
| FS | Com | 13 06.4 | +22 37 | 5.3–6.1 | SR | 58: | M |
| R | CVn | 13 49.0 | +39 33 | 6.5–12.9 | M | 328.53 | M |
| ZZ | Boo | 13 56.2 | +25 55 | 5.8–6.4 | EW | 4.99 | G |
| V | Boo | 14 29.8 | +38 52 | 7.0–12.0 | SR | 258 | M |
| R | Boo | 14 37.2 | +26 44 | 6.2–13.1 | M | 233.4 | M |
| W | Boo | 14 43.4 | +26 32 | 4.7–5.4 | SR | 450: | M |
| 44 | Boo | 15 03.8 | +47 39 | 5.8–6.4 | EW | 0.27 | G + G i Boo |
| S | CrB | 15 21.4 | +31 22 | 5.8–14.1 | M | 360.3 | M |
| R | CrB | 15 48.6 | +28 09 | 5.7–14.8 | RCB | — | G |
| T | CrB | 15 59.5 | +25 55 | 2.0–10.8 | RN | 29,000: | M |

## Double stars

| Star | Con | RA h m | Dec ° ′ | PA ° | Sep ″ | Magnitudes | |
|---|---|---|---|---|---|---|---|
| 2 | CVn | 12 16.1 | +40 40 | 260 | 11.4 | 5.8 + 8.1 | |
| α | CVn | 12 56.0 | +38 19 | 229 | 19.4 | 2.9 + 5.5 | Cor Caroli |
| 78 | UMa | 13 00.7 | +56 22 | 69 | 1.5 | 5.0 + 7.4 | Binary, 115.7 years |
| ζ | UMa | 13 23.9 | +54 56 | 152 | 14.4 | 2.3 + 4.0 | Mizar |
| | | | | 71 | 708.7 | 4.0 | 80 UMa, Alcor |
| κ | Boo | 14 13.5 | +51 47 | 236 | 13.4 | 4.6 + 6.6 | |
| ε | Boo | 14 45.0 | +27 04 | 339 | 2.8 | 2.5 + 4.9 | Izar |
| 39 | Boo | 14 49.7 | +48 43 | 45 | 2.9 | 6.2 + 6.9 | |
| 44 | Boo | 15 03.8 | +47 39 | 53 | 2.2 | 5.3 + 6.2 v | Binary, 225 years; variable |
| η | CrB | 15 23.2 | +30 17 | 63 | 0.8 | 5.6 + 5.9 | Binary, 41.6 years |
| μ¹, μ² | Boo | 15 24.5 | +37 23 | 171 | 108.3 | 4.3 + 6.5 d | Alkalurops |
| μ² | Boo | 15 24.5 | +37 21 | 8 | 2.3 | 7.0 + 7.6 | Binary, 260.1 years |

## Globular clusters

| NGC/IC | Other | Con | RA h m | Dec ° ′ | Mag | Diam ′ |
|---|---|---|---|---|---|---|
| 5272 | M3 | CVn | 13 42.2 | +28 23 | 6.4 | 16.2 |
| 5466 | | Boo | 14 05.5 | +28 32 | 9.1 | 11.0 |

## Galaxies

| NGC/IC | Other | Con | RA h m | Dec ° ′ | Mag | Size ′ | Type | |
|---|---|---|---|---|---|---|---|---|
| 4036 | | UMa | 12 01.4 | +61 54 | 10.6 | 4.5 × 2.0 | E6 | |
| 4041 | | UMa | 12 02.2 | +62 08 | 11.1 | 2.8 × 2.7 | Sc | |
| 4051 | | UMa | 12 03.2 | +44 32 | 10.3 | 5.0 × 4.0 | Sc | |
| 4062 | | UMa | 12 04.1 | +31 54 | 11.2 | 4.3 × 2.0 | Sb | |
| 4088 | | UMa | 12 05.6 | +50 33 | 10.5 | 5.8 × 2.5 | Sc | |
| 4096 | | UMa | 12 06.0 | +47 29 | 10.6 | 6.5 × 2.0 | Sc | |
| 4102 | | UMa | 12 06.4 | +52 43 | 12.3 | 3.2 × 1.9 | Sc | |
| 4111 | | CVn | 12 07.1 | +43 04 | 10.8 | 4.8 × 1.1 | S0 | |

## Galaxies *(continued)*

| NGC/IC | Other | Con | RA h m | Dec ° ′ | Mag | Size ′ | Type | |
|---|---|---|---|---|---|---|---|---|
| 4136 | | Com | 12 09.3 | +29 56 | 11.4 | 4.1 × 3.9 | Sc | |
| 4143 | | CVn | 12 09.6 | +42 32 | 12.1 | 2.9 × 1.8 | E4 | |
| 4145 | | CVn | 12 10.0 | +39 53 | 11.0 | 5.8 × 4.4 | Sc | |
| 4151 | | CVn | 12 10.5 | +39 24 | 10.4 | 5.9 × 4.4 | Pec | |
| 4157 | | UMa | 12 11.1 | +50 29 | 11.7 | 6.9 × 1.7 | Sb | |
| 4214 | | CVn | 12 15.6 | +36 20 | 9.7 | 7.9 × 6.3 | Irr | |
| 4217 | | CVn | 12 15.8 | +47 06 | 11.9 | 5.5 × 1.8 | Sb | |
| 4242 | | CVn | 12 17.5 | +45 37 | 11.0 | 4.8 × 3.8 | S | |
| 4244 | | CVn | 12 17.5 | +37 49 | 10.2 | 16.2 × 2.5 | S | |
| 4251 | | Com | 12 18.1 | +28 10 | 11.6 | 4.2 × 1.9 | E7 | |
| 4258 | M106 | CVn | 12 19.0 | +47 18 | 8.3 | 18.2 × 7.9 | Sb | |
| 4274 | | Com | 12 19.8 | +29 37 | 10.4 | 6.9 × 2.8 | Sb | |
| 4278 | | Com | 12 20.1 | +29 17 | 10.2 | 3.6 × 3.5 | E1 | |
| 4314 | | Com | 12 22.6 | +29 53 | 10.5 | 4.8 × 4.3 | SBa | |
| 4395 | | CVn | 12 25.8 | +33 33 | 10.2 | 12.9 × 11.0 | S | |
| 4414 | | Com | 12 26.4 | +31 13 | 10.3 | 3.6 × 2.2 | Sc | |
| 4448 | | Com | 12 28.2 | +28 37 | 11.1 | 4.0 × 1.6 | Sb | |
| 4449 | | CVn | 12 28.2 | +44 06 | 9.4 | 5.1 × 3.7 | Irr | |
| 4490 | | Com | 12 30.6 | +41 38 | 9.8 | 5.9 × 3.1 | Sc | |
| 4494 | | Com | 12 31.4 | +25 47 | 9.7 | 4.8 × 3.8 | E1 | |
| 4559 | | Com | 12 36.0 | +27 58 | 9.9 | 10.5 × 4.9 | Sc | |
| 4565 | | Com | 12 36.3 | +25 59 | 9.6 | 16.2 × 2.8 | SBc | |
| 4605 | | UMa | 12 40.0 | +61 37 | 11.0 | 5.5 × 2.3 | Sc | |
| 4618 | | CVn | 12 41.5 | +41 09 | 10.8 | 4.4 × 3.8 | Sc | |
| 4631 | | CVn | 12 42.1 | +32 32 | 9.3 | 15.1 × 3.3 | Sc | |
| 4656 | | CVn | 12 44.0 | +32 10 | 10.4 | 13.8 × 3.3 | Sc | |
| 4725 | | Com | 12 50.4 | +25 30 | 9.2 | 11.0 × 7.9 | SBb | |
| 4736 | M94 | CVn | 12 50.9 | +41 07 | 8.2 | 11.0 × 9.1 | Sb | |
| 4800 | | CVn | 12 54.6 | +46 32 | 12.3 | 1.8 × 1.4 | Sb | |
| 4826 | M64 | Com | 12 56.7 | +21 41 | 8.5 | 9.3 × 5.4 | Sb | Black-Eye Galaxy |
| 5005 | | CVn | 13 10.9 | +37 03 | 9.8 | 5.4 × 2.7 | Sb | |
| 5033 | | CVn | 13 13.4 | +36 36 | 10.1 | 10.5 × 5.6 | Sb | |
| 5055 | M63 | CVn | 13 15.8 | +42 02 | 8.6 | 12.3 × 7.6 | Sb | Sunflower Galaxy |
| 5112 | | CVn | 13 21.9 | +38 44 | 11.9 | 3.9 × 2.9 | Sc | |
| 5204 | | UMa | 13 29.6 | +58 25 | 11.3 | 4.8 × 3.0 | Irr | |
| 5194 | M51 | CVn | 13 29.9 | +47 12 | 8.4 | 11.0 × 7.8 | Sc | Whirlpool Galaxy |
| 5195 | | CVn | 13 30.0 | +47 16 | 9.6 | 5.4 × 4.3 | Pec | Companion of M51 |
| 5322 | | UMa | 13 49.3 | +60 12 | 10.0 | 5.5 × 3.9 | E2 | |
| 5371 | | CVn | 13 55.7 | +40 28 | 10.8 | 4.4 × 3.6 | Sb | |
| 5377 | | CVn | 13 56.3 | +47 14 | 11.2 | 4.6 × 2.7 | Sa | |
| 5457 | M101 | UMa | 14 03.2 | +54 21 | 7.7 | 26.9 × 26.3 | Sc | Pinwheel Galaxy |
| 5474 | | UMa | 14 05.0 | +53 40 | 10.9 | 4.5 × 4.2 | Sc | |
| 5585 | | UMa | 14 19.8 | +56 44 | 10.9 | 5.5 × 3.7 | S | |
| 5676 | | Boo | 14 32.8 | +49 28 | 10.9 | 3.9 × 2.0 | Sc | |
| 5866 | (M102) | Dra | 15 06.5 | +55 46 | 10.0 | 5.2 × 2.3 | E6 | Spindle Galaxy |
| 5879 | | Dra | 15 09.8 | +57 00 | 11.5 | 4.4 × 1.7 | Sb | |
| 5907 | | Dra | 15 15.9 | +56 19 | 10.4 | 12.3 × 1.8 | Sb | |
| 5982 | | Dra | 15 38.7 | +59 21 | 11.1 | 2.9 × 2.2 | E3 | |
| 5985 | | Dra | 15 39.6 | +59 20 | 11.0 | 5.5 × 3.2 | Sb | |
| 6015 | | Dra | 15 51.4 | +62 19 | 11.2 | 5.4 × 2.3 | Sc | |

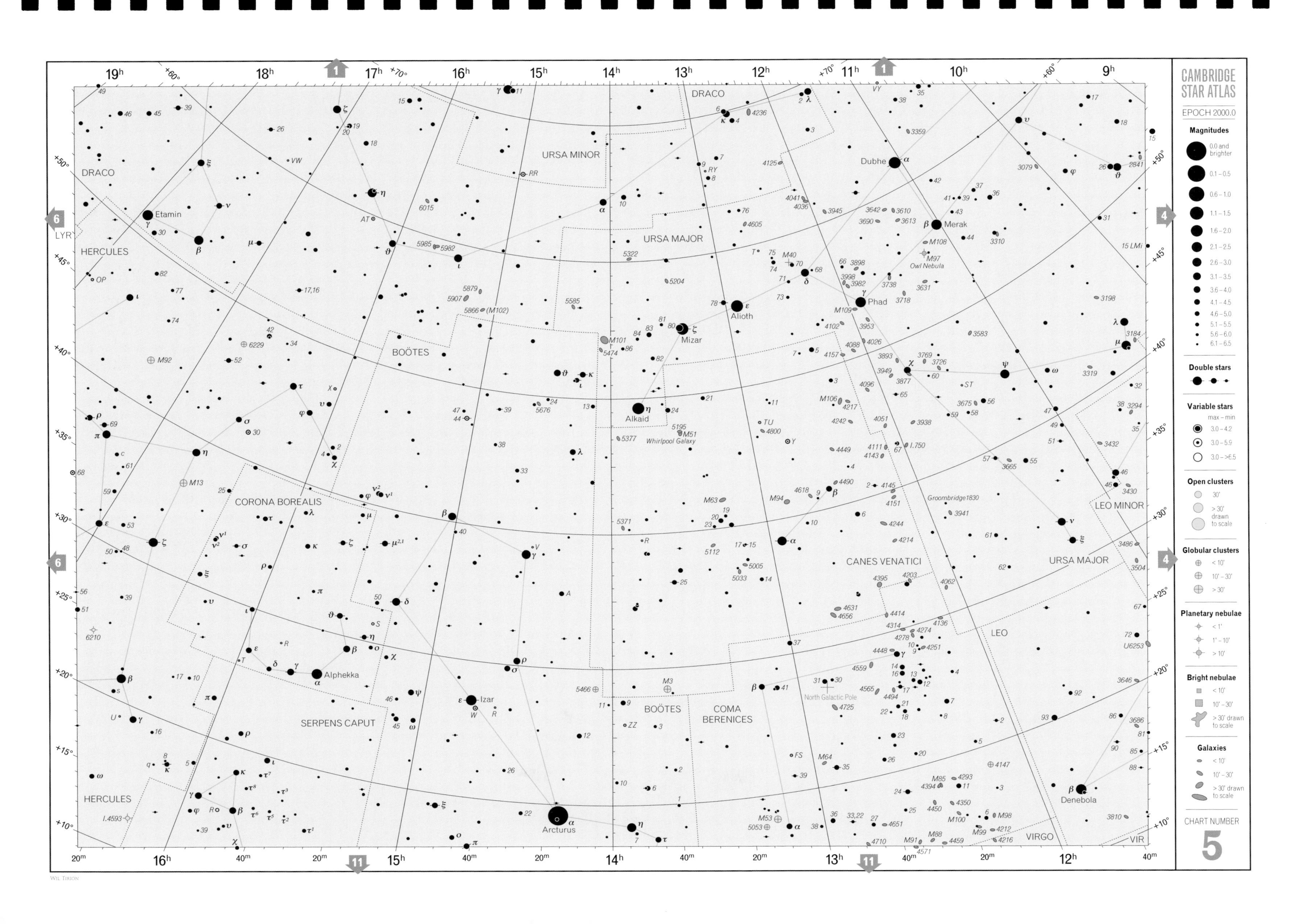

CAMBRIDGE STAR ATLAS
EPOCH 2000.0
Magnitudes
0.0 and brighter
0.1 – 0.5
0.6 – 1.0
1.1 – 1.5
1.6 – 2.0
2.1 – 2.5
2.6 – 3.0
3.1 – 3.5
3.6 – 4.0
4.1 – 4.5
4.6 – 5.0
5.1 – 5.5
5.6 – 6.0
6.1 – 6.5
Double stars
Variable stars
max – min
3.0 – 4.2
3.0 – 5.9
3.0 – >6.5
Open clusters
30'
> 30' drawn to scale
Globular clusters
< 10'
10' – 30'
> 30'
Planetary nebulae
< 1'
1' – 10'
> 10'
Bright nebulae
< 10'
10' – 30'
> 30' drawn to scale
Galaxies
< 10'
10' – 30'
> 30' drawn to scale
CHART NUMBER
5
DRACO
URSA MINOR
URSA MAJOR
LEO MINOR
LEO
VIRGO
VIR
CANES VENATICI
COMA BERENICES
BOÖTES
CORONA BOREALIS
SERPENS CAPUT
HERCULES
LYR
Dubhe
Merak
Phad
Alioth
Mizar
Alkaid
Denebola
Arcturus
Izar
Alphekka
Etamin
Owl Nebula
Whirlpool Galaxy
North Galactic Pole
Groombridge1830
WIL TIRION

## Variable stars

| Star | Con | RA h m | Dec ° ′ | Range | Type | Period (days) | Spectrum | |
|---|---|---|---|---|---|---|---|---|
| X | Her | 16 02.7 | +47 14 | 6.3–7.4 | SR | 95.0 | M | |
| AT | Dra | 16 17.3 | +59 45 | 5.3–6.0 | Irr | — | M | |
| 30 | Her | 16 28.6 | +41 53 | 4.3–6.3 | SR | 70: | M | g Her |
| VW | Dra | 17 16.5 | +60 40 | 6.0–6.5 | SR | 170: | K | |
| 68 | Her | 17 17.3 | +33 06 | 4.7–5.4 | EB | 2.05 | B | u Her |
| OP | Her | 17 56.8 | +45 21 | 5.9–6.7 | Irr | — | M | |
| XY | Lyr | 18 38.1 | +39 40 | 5.8–6.4 | Irr | — | M | |
| β | Lyr | 18 50.1 | +33 22 | 3.3–4.3 | EB | 12.94 | B + A | Sheliak |
| R | Lyr | 18 55.3 | +43 57 | 3.9–5.0 | SR | 46.0 | M | |
| AF | Cyg | 19 30.2 | +46 09 | 6.4–8.4 | SR | 94.1 | M | |
| U | Vul | 19 36.6 | +20 20 | 6.8–7.5 | Cep | 7.99 | F – G | |
| R | Cyg | 19 36.8 | +50 12 | 6.1–14.2 | M | 426.4 | S | |
| V1143 | Cyg | 19 38.7 | +54 58 | 5.9–6.4 | EA | 7.64 | F | |
| RT | Cyg | 19 43.6 | +48 47 | 6.4–12.7 | M | 190.3 | M | |
| V973 | Cyg | 19 44.8 | +40 43 | 6.1–6.6 | SR | 40: | M | |
| SU | Cyg | 19 44.8 | +29 16 | 6.5–7.2 | Cep | 3.85 | F | |
| χ | Cyg | 19 50.6 | +32 55 | 3.3–14.2 | M | 406.9 | S | |
| V449 | Cyg | 19 53.3 | +33 57 | 7.4–9.0 p | Irr | — | M | |

## Double stars

| Star | Con | RA h m | Dec ° ′ | PA ° | Sep ″ | Magnitudes | |
|---|---|---|---|---|---|---|---|
| η | Dra | 16 24.0 | +61 31 | 142 | 5.2 | 2.7 + 8.7 | |
| 17 | Dra | 16 36.2 | +52 55 | 108 | 3.4 | 5.4 + 6.4 | |
| | | | | 194 | 90.3 | 5.5 | 16 Dra |
| ζ | Her | 16 41.3 | +31 36 | 12 | 0.8 | 2.9 + 5.5 | Binary, 34.5 years |
| μ | Dra | 17 05.3 | +54 28 | 8 | 1.9 | 5.7 + 5.7 | Binary, 482 years |
| ρ | Her | 17 23.7 | +37 09 | 316 | 4.1 | 4.6 + 5.6 | |
| ν | Dra | 17 32.2 | +55 11 | 312 | 61.9 | 4.9 + 4.9 | |
| μ | Her | 17 46.5 | +27 43 | 247 | 33.8 | 3.4 + 10.1 | |
| 90 | Her | 17 53.3 | +40 00 | 116 | 1.6 | 5.2 + 8.5 | |
| 95 | Her | 18 01.5 | +21 36 | 258 | 6.3 | 5.0 + 5.1 | |
| 100 | Her | 18 07.8 | +26 06 | 183 | 14.2 | 5.9 + 6.0 | |
| 39 | Dra | 18 23.9 | +58 48 | 351 | 3.1 | 5.0 + 8.0 | |
| | | | | 21 | 88.9 | 7.4 | 5 other fainter components |
| ε | Lyr | 18 44.3 | +39 40 | 207.7 | 173 | 4.7 + 4.6 | ε¹, ε² |
| | | | | 350 | 2.6 | 5.0 + 6.1 | ε¹, binary, 1165 years |
| | | | | 82 | 2.3 | 5.2 + 5.5 | ε², binary, 585 years |
| ζ | Lyr | 18 44.8 | +37 36 | 150 | 43.7 | 4.3 + 5.9 | ζ¹, ζ² |
| β | Lyr | 18 50.1 | +33 22 | 149 | 45.7 | 3.4 v + 8.6 | Sheliak; variable |
| ο | Dra | 18 51.2 | +59 23 | 326 | 34.2 | 4.8 + 7.8 | |
| η | Lyr | 19 13.8 | +39 09 | 82 | 28.1 | 4.4 + 9.1 | |
| 2 | Vul | 19 17.7 | +23 02 | 127 | 1.8 | 5.4 + 9.2 | ES Vul |
| α, 8 | Vul | 19 28.7 | +24 40 | 28 | 413.7 | 4.4 + 5.8 | |
| β | Cyg | 19 30.9 | +27 58 | 54 | 34.4 | 3.1 + 5.1 | Albireo |
| δ | Cyg | 19 45.0 | +45 08 | 221 | 2.5 | 2.9 + 6.3 | Binary, 827.6 years |
| 17 | Cyg | 19 46.4 | +33 44 | 69 | 26.0 | 5.0 + 9.2 | |
| ψ | Cyg | 19 55.6 | +52 26 | 178 | 3.2 | 4.9 + 7.4 | |

## Open clusters

| NGC/IC | Other | Con | RA h m | Dec ° ′ | Mag | Diam ′ | N* | |
|---|---|---|---|---|---|---|---|---|
| — | Steph 1 | Lyr | 18 53.5 | +36 55 | 3.8 | 20 | 15 | δ Lyrae Cluster |
| — | Cr 399 | Vul | 19 25.4 | +20 11 | 3.6 | 60 | 40 | Brocchi's Cluster |
| 6811 | | Cyg | 19 38.2 | +46 34 | 6.8 | 13 | 70 | |
| 6819 | | Cyg | 19 41.3 | +40 11 | 7.3 | 5 | | |
| 6823 | | Vul | 19 43.1 | +23 18 | 7.1 | 12 | 30 | In nebula 6820 |
| 6830 | | Vul | 19 51.0 | +23 04 | 7.9 | 12 | 20 | |
| 6834 | | Cyg | 19 52.2 | +29 25 | 7.8 | 5 | 50 | |

## Globular clusters

| NGC/IC | Other | Con | RA h m | Dec ° ′ | Mag | Diam ′ | |
|---|---|---|---|---|---|---|---|
| 6205 | M13 | Her | 16 41.7 | +36 28 | 5.9 | 16.6 | Hercules Cluster |
| 6229 | | Her | 16 47.0 | +47 32 | 9.4 | 4.5 | |
| 6341 | M92 | Her | 17 17.1 | +43 08 | 6.5 | 11.2 | |
| 6779 | M56 | Lyr | 19 16.6 | +30 11 | 8.2 | 7.1 | |

## Planetary nebulae

| NGC/IC | Other | Con | RA h m | Dec ° ′ | Mag p | Diam ″ | Mag* | |
|---|---|---|---|---|---|---|---|---|
| 6210 | | Her | 16 44.5 | +23 49 | 9.3 | 14 | 12.9 | |
| — | PK51+9.1 | Her | 18 49.7 | +20 51 | 12.2 | 3 | 13.0 | |
| 6720 | M57 | Lyr | 18 53.6 | +33 02 | 9.7 | 70/150 | 14.8 | Ring Nebula |
| — | PK64+5.1 | Cyg | 19 34.8 | +30 31 | 9.6 | 8 | 10.0 | |
| 6826 | | Cyg | 19 44.8 | +50 31 | 9.8 | 30/140 | 10.4 | Blinking Planetary |
| 6853 | M27 | Vul | 19 59.6 | +22 43 | 7.6 | 350/910 | 13.9 | Dumbbell Nebula |

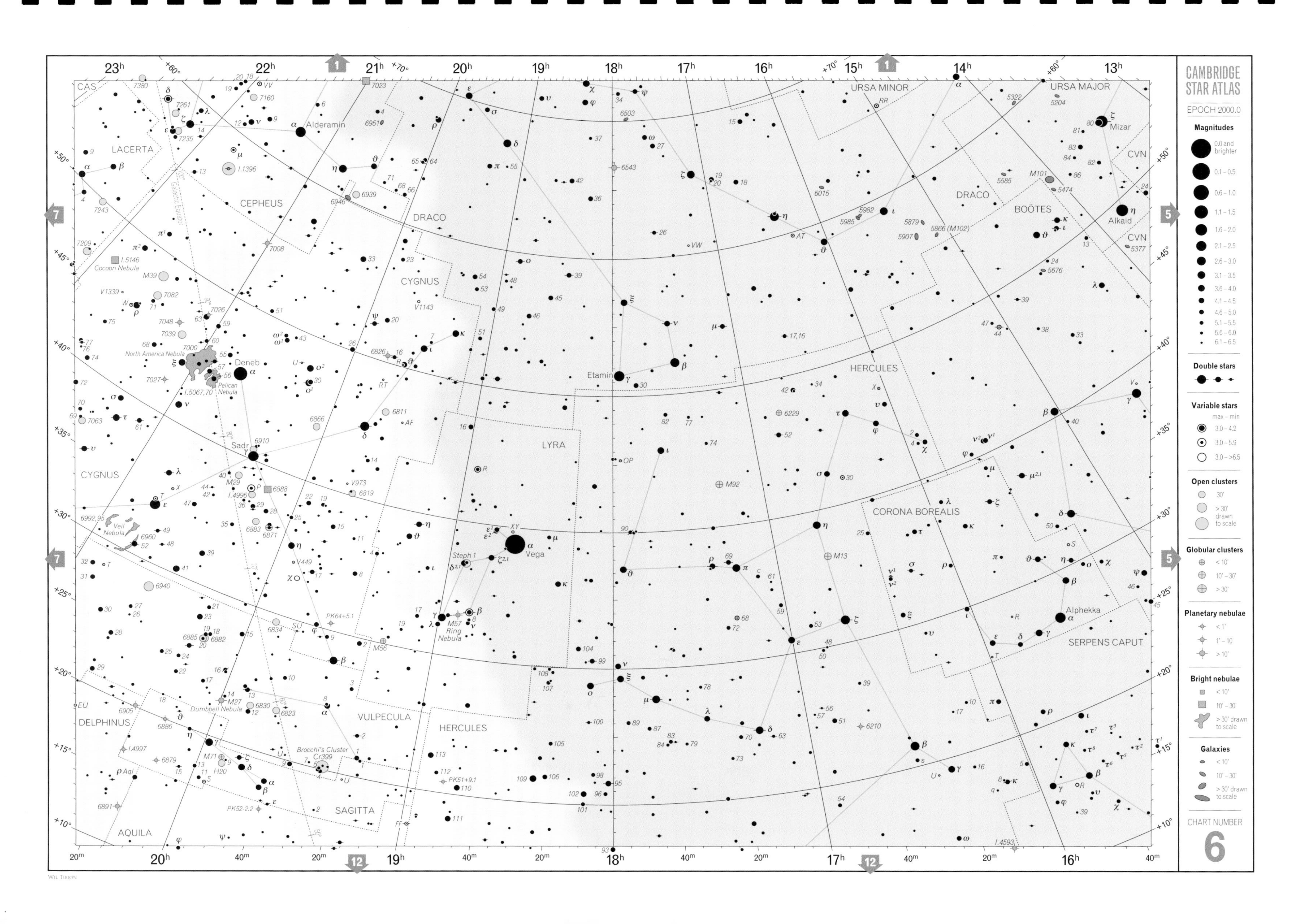
CAMBRIDGE STAR ATLAS
EPOCH 2000.0
Magnitudes
0.0 and brighter
0.1 – 0.5
0.6 – 1.0
1.1 – 1.5
1.6 – 2.0
2.1 – 2.5
2.6 – 3.0
3.1 – 3.5
3.6 – 4.0
4.1 – 4.5
4.6 – 5.0
5.1 – 5.5
5.6 – 6.0
6.1 – 6.5
Double stars
Variable stars
max – min
3.0 – 4.2
3.0 – 5.9
3.0 – >6.5
Open clusters
30'
> 30' drawn to scale
Globular clusters
< 10'
10' – 30'
> 30'
Planetary nebulae
< 1'
1' – 10'
> 10'
Bright nebulae
< 10'
10' – 30'
> 30' drawn to scale
Galaxies
< 10'
10' – 30'
> 30' drawn to scale
CHART NUMBER
6
URSA MAJOR
URSA MINOR
CVN
BOÖTES
DRACO
HERCULES
CORONA BOREALIS
SERPENS CAPUT
LYRA
CYGNUS
CEPHEUS
LACERTA
CAS
VULPECULA
SAGITTA
DELPHINUS
AQUILA
Mizar
Alkaid
Alphekka
Etamin
Vega
Deneb
Sadr
Alderamin
North America Nebula
Pelican Nebula
Cocoon Nebula
Veil Nebula
M57 Ring Nebula
Dumbbell Nebula
Brocchi's Cluster
Galactic Equator
Wil Tirion

# Chart 7 *RA $20^h$ to $0^h$, declination +65° to +20°*

## Variable stars

| Star | Con | RA h m | Dec ° ′ | Range | Type | Period (days) | Spectrum |
|---|---|---|---|---|---|---|---|
| P | Cyg | 20 17.8 | +38 02 | 3.0–6.0 | SD | — | B |
| U | Cyg | 20 19.6 | +47 54 | 5.9–12.1 | M | 462.4 | M |
| X | Cyg | 20 43.4 | +35 35 | 5.9–6.7 | Cep | 16.39 | F–G |
| T | Cyg | 20 47.2 | +34 22 | 5.0–5.5 | Irr | — | K |
| T | Vul | 20 51.5 | +28 15 | 5.4–6.1 | Cep | 4.43 | F–G |
| W | Cyg | 21 36.0 | +45 22 | 5.0–7.6 | SR | 126.3 | M |
| V460 | Cyg | 21 42.0 | +35 31 | 5.6–7.0 | Irr | — | M |
| V1339 | Cyg | 21 42.1 | +45 46 | 5.9–7.1 | SR | 35: | M |
| μ | Cep | 21 43.5 | +58 47 | 3.4–5.1 | SR | 730 | M |
| VV | Cep | 21 56.7 | +63 38 | 4.8–5.4 | EA | 7430 | M |
| AR | Lac | 22 08.7 | +45 45 | 6.1–6.8 | EA | 1.98 | G+K |
| δ | Cep | 22 29.2 | +58 25 | 3.9–4.4 | Cep | 5.37 | F–G |
| V509 | Cas | 23 00.1 | +56 57 | 4.8–5.5 | SR | — | F–K |
| β | Peg | 23 03.8 | +28 05 | 2.3–2.7 | Irr | — | M Scheat |
| Z | And | 23 33.7 | +48 49 | 8.0–12.4 p | ZA | — | M |
| ρ | Cas | 23 54.4 | +57 30 | 4.1–6.2 | SR | 320 | F–K |
| R | Cas | 23 58.4 | +51 24 | 4.7–13.5 | M | 430.5 | M |

## Double stars

| Star | Con | RA h m | Dec ° ′ | PA ° | Sep ″ | Magnitudes | |
|---|---|---|---|---|---|---|---|
| 52 | Cyg | 20 45.7 | +30 43 | 67 | 6.0 | 4.2 + 6.4 | In Veil Nebula 6920 |
| 61 | Cyg | 21 06.9 | +38 45 | 150 | 30.3 | 5.2 + 6.0 | Binary, 653.3 years; large p.m. |
| τ | Cyg | 21 14.8 | +38 03 | 306 | 0.8 | 3.8 + 6.4 | Binary, 49.9 years |
| υ | Cyg | 21 17.9 | +34 54 | 220 | 15.1 | 4.4 + 10.0 | |
| μ | Cyg | 21 44.1 | +28 45 | 320 | 1.2 | 4.8 + 6.1 | Binary, 507.5 years |
| ξ | Cep | 22 03.8 | +64 38 | 274 | 8.2 | 4.4 + 6.5 | Binary, 3800 years |
| δ | Cep | 22 29.2 | +58 25 | 191 | 41.0 | 3.9 v + 7.5 | Variable |
| 8 | Lac | 22 35.9 | +39 38 | 186 | 22.4 | 5.7 + 6.5 | |
| | | | | 169 | 48.8 | 10.5 | |
| | | | | 144 | 81.8 | 9.3 | |
| | | | | 239 | 336.6 | 7.8 | |
| 72 | Peg | 23 34.0 | +31 20 | 97 | 0.5 | 5.7 + 5.8 | Binary, 241.2 years |
| 78 | Peg | 23 44.0 | +29 22 | 235 | 1.0 | 5.0 + 8.1 | |
| 6 | Cas | 23 48.8 | +62 13 | 193 | 1.6 | 5.5 + 8.0 | |
| σ | Cas | 23 59.0 | +55 45 | 326 | 3.0 | 5.0 + 7.1 | |

## Open clusters

| NGC/IC | Other | Con | RA h m | Dec ° ′ | Mag | Diam ′ | N* | |
|---|---|---|---|---|---|---|---|---|
| 6866 | | Cyg | 20 03.7 | +44 00 | 7.6 | 7 | 80 | |
| 6871 | | Cyg | 20 05.9 | +35 47 | 5.2 | 20 | 15 | |
| 6883 | | Cyg | 20 11.3 | +35 51 | 8.0 | 15 | 30 | |
| 6882 | | Vul | 20 11.7 | +26 33 | 8.1 | 18 | | Near cluster 6885 |
| 6885 | | Vul | 20 12.0 | +26 29 | 5.7 | 7 | 30 | Contains 20 Vul |
| I.4996 | | Cyg | 20 16.5 | +37 38 | 7.3 | 6 | 15 | |
| 6910 | | Cyg | 20 23.1 | +40 47 | 7.4 | 8 | 50 | |
| 6913 | M29 | Cyg | 20 23.9 | +38 32 | 6.6 | 7 | 50 | |
| 6939 | | Cep | 20 31.4 | +60 38 | 7.8 | 8 | 80 | |
| 6940 | | Vul | 20 34.6 | +28 18 | 6.3 | 31 | 60 | |
| 7039 | | Cyg | 21 11.2 | +45 39 | 7.6 | 25 | 50 | |

## Open clusters *(continued)*

| NGC/IC | Other | Con | RA h m | Dec ° ′ | Mag | Diam ′ | N* |
|---|---|---|---|---|---|---|---|
| 7063 | | Cyg | 21 24.4 | +36 30 | 7.0 | 8 | 12 |
| 7082 | | Cyg | 21 29.4 | +47 05 | 7.2 | 25 | |
| 7092 | M39 | Cyg | 21 32.2 | +48 26 | 4.6 | 32 | 30 |
| I.1396 | | Cep | 21 39.1 | +57 30 | 3.5 | 50 | 50 |
| 7160 | | Cep | 21 53.7 | +62 36 | 6.1 | 7 | 12 |
| 7209 | | Lac | 22 05.2 | +46 30 | 6.7 | 25 | 25 |
| 7235 | | Cep | 22 12.6 | +57 17 | 7.7 | 4 | 30 |
| 7243 | | Lac | 22 15.3 | +49 53 | 6.4 | 21 | 40 |
| 7261 | | Cep | 22 20.4 | +58 05 | 8.4 | 6 | 30 |
| 7380 | | Cep | 22 47.0 | +58 06 | 7.2 | 12 | 40 |
| 7510 | | Cep | 23 11.5 | +60 34 | 7.9 | 4 | 60 |
| 7654 | M52 | Cas | 23 24.2 | +61 35 | 6.9 | 13 | 100 |
| 7686 | | And | 23 30.2 | +49 08 | 5.6 | 15 | 20 |
| 7789 | | Cas | 23 57.0 | +56 44 | 6.7 | 13 | 300 |
| 7790 | | Cas | 23 58.4 | +61 13 | 8.5 | 17 | 40 |

## Bright diffuse nebulae

| NGC/IC | Other | Con | RA h m | Dec ° ′ | Type | Diam ′ | Mag* | |
|---|---|---|---|---|---|---|---|---|
| 6888 | | Cyg | 20 12.0 | +38 21 | E | 20 × 10 | 7.4 | |
| 6960 | | Cyg | 20 45.7 | +30 43 | E | 70 × 6 | | Veil Nebula SNR |
| | | | 20 48.5 | +31 09 | E | 45 × 30 | | Veil Nebula SNR |
| I.5067, I.5070 | | Cyg | 20 50.8 | +44 21 | E | 80 × 70 | | Pelican Nebula |
| 6992 | | Cyg | 20 56.4 | +31 43 | E | 60 × 8 | | Veil Nebula SNR |
| 6995 | | Cyg | 20 57.1 | +31 13 | E | 12 | | Veil Nebula SNR |
| 7000 | | Cyg | 20 58.8 | +44 20 | E | 120 × 100 | 6.0 | North America Nebula |
| I.5146 | | Cyg | 21 53.5 | +47 16 | E | 12 × 12 | 10.0 | Cocoon Nebula |
| 7635 | | Cas | 23 20.7 | +61 12 | E | 15 × 8 | 6.9 | Bubble Nebula |

## Planetary nebulae

| NGC/IC | Other | Con | RA h m | Dec ° ′ | Mag p | Diam ″ | Mag* |
|---|---|---|---|---|---|---|---|
| 6905 | | Del | 20 22.4 | +20 07 | 11.9 | 46/100 | 13.5 |
| 7008 | | Cyg | 21 00.6 | +54 33 | 13.3 | 83 | 13.2 |
| 7026 | | Cyg | 21 06.3 | +47 51 | 12.7 | 21 | 14.5 |
| 7027 | | Cyg | 21 07.1 | +42 14 | 10.4 | 15 | 11.3 |
| 7048 | | Cyg | 21 14.2 | +46 16 | 11.3 | 61 | 18.3 |
| 7662 | | And | 23 25.9 | +42 33 | 9.2 | 20/130 | 13.2 |

## Galaxies

| NGC/IC | Other | Con | RA h m | Dec ° ′ | Mag | Size ′ | Type |
|---|---|---|---|---|---|---|---|
| 6946 | | Cep | 20 34.8 | +60 09 | 8.9 | 11.0 × 9.8 | Sc |
| 7217 | | Peg | 22 07.9 | +31 22 | 10.2 | 3.7 × 3.2 | Sb |
| 7331 | | Peg | 22 37.1 | +34 25 | 9.5 | 10.7 × 4.0 | Sb |
| 7332 | | Peg | 22 37.4 | +23 48 | 11.0 | 4.2 × 1.3 | E7 |
| 7457 | | Peg | 23 01.0 | +30 09 | 10.8 | 4.4 × 2.5 | E |
| 7640 | | And | 23 22.1 | +40 51 | 10.9 | 10.7 × 2.5 | SBb |
| 7741 | | Peg | 23 43.9 | +26 05 | 11.4 | 4.0 × 2.8 | SBc |

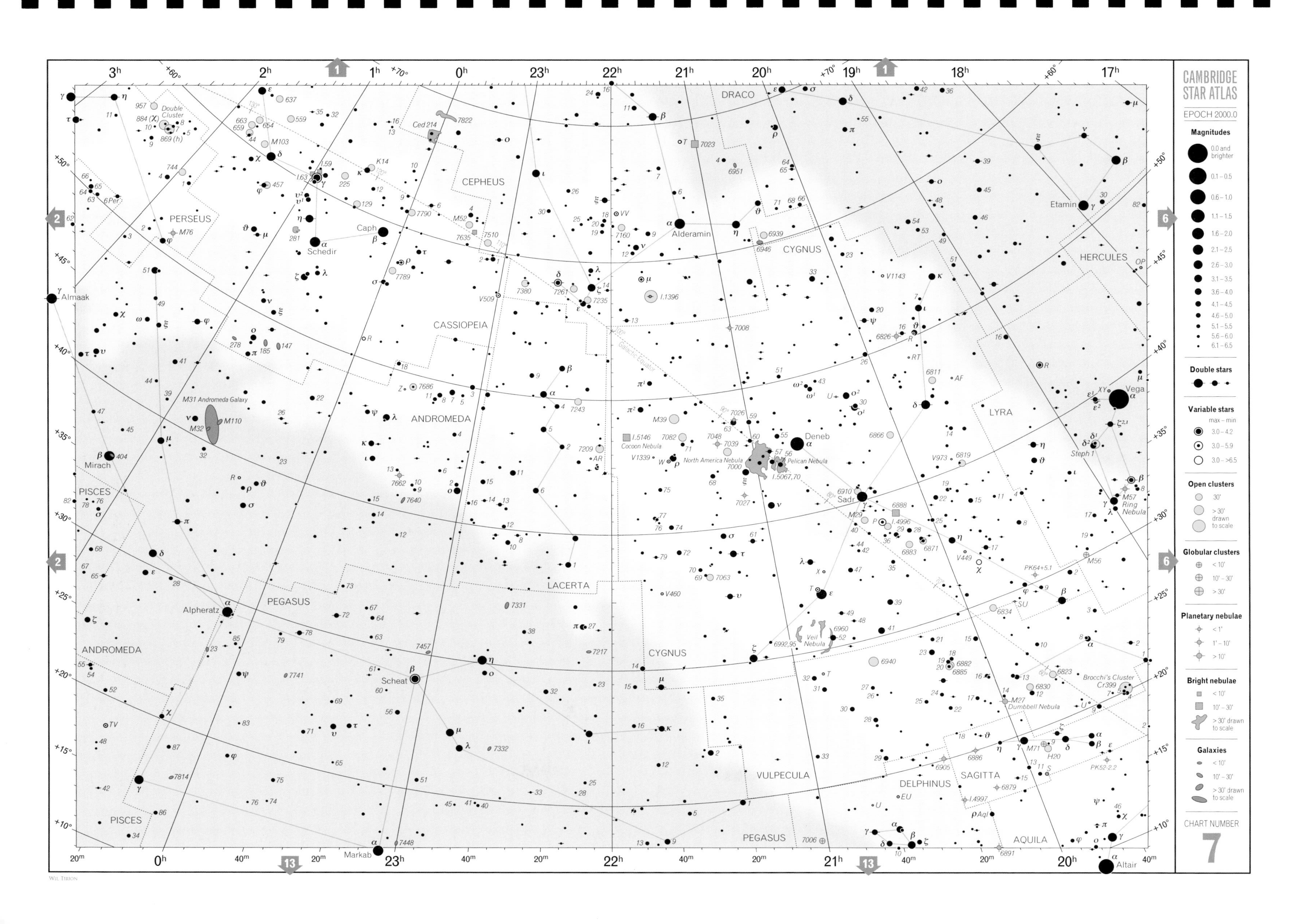

CAMBRIDGE STAR ATLAS
EPOCH 2000.0
Magnitudes
Double stars
Variable stars
Open clusters
Globular clusters
Planetary nebulae
Bright nebulae
Galaxies
CHART NUMBER
7
HERCULES
LYRA
DRACO
CYGNUS
CEPHEUS
CASSIOPEIA
ANDROMEDA
LACERTA
PEGASUS
PERSEUS
PISCES
VULPECULA
SAGITTA
DELPHINUS
AQUILA
Vega
Deneb
Sadr
Alderamin
Etamin
Caph
Schedir
Scheat
Markab
Alpheratz
Mirach
Almaak
Altair
M57 Ring Nebula
M27 Dumbbell Nebula
Brocchi's Cluster
Veil Nebula
North America Nebula
Pelican Nebula
Cocoon Nebula
M31 Andromeda Galaxy
M110
M32
Double Cluster
Galactic Equator
WIL TIRION

## Variable stars

| Star | Con | RA h m | Dec ° ′ | Range | Type | Period (days) | Spectrum |
|---|---|---|---|---|---|---|---|
| TV | Psc | 00 28.0 | +17 54 | 4.7–5.4 | SR | 70 | M |
| o | Cet | 02 19.3 | −02 59 | 2.0–10.1 | M | 332.0 | M Mira |
| U | Cet | 02 33.7 | −13 09 | 6.8–13.4 | M | 234.8 | M |
| Z | Eri | 02 47.9 | −12 28 | 5.6–7.2 | SR | 80 | M |
| RR | Eri | 02 52.2 | −08 16 | 6.3–8.1 | SR | 97 | M |

## Double stars

| Star | Con | RA h m | Dec ° ′ | PA ° | Sep ″ | Magnitudes | |
|---|---|---|---|---|---|---|---|
| 51 | Psc | 00 32.4 | +06 57 | 83 | 27.5 | 5.7 + 9.5 | |
| 26 | Cet | 01 03.8 | +01 22 | 253 | 16.0 | 6.2 + 8.6 | |
| ζ | Psc | 01 13.7 | +07 35 | 63 | 23.0 | 5.6 + 6.5 | |
| 37 | Cet | 01 14.4 | −07 55 | 331 | 49.7 | 5.2 + 8.7 | |
| χ | Cet | 01 49.6 | −10 41 | 250 | 183.8 | 4.9 + 6.9 | |
| γ | Ari | 01 53.5 | +19 18 | 0 | 7.8 | 4.8 + 4.8 | Mesartim |
| α | Psc | 02 02.0 | +02 46 | 272 | 1.8 | 4.2 + 5.1 | Binary, 933.1 years |
| 66 | Cet | 02 12.8 | −02 24 | 234 | 16.5 | 5.7 + 7.5 | |
| 84 | Cet | 02 41.2 | −02 42 | 310 | 4.0 | 5.8 + 9.0 | |
| γ | Cet | 02 43.3 | +03 14 | 294 | 2.8 | 3.5 + 7.3 | |
| π | Ari | 02 49.3 | +17 28 | 120 | 3.2 | 5.2 + 8.7 | |
| $\rho^2$ | Eri | 03 02.7 | −07 41 | 75 | 1.8 | 5.3 + 9.5 | |
| 95 | Cet | 03 18.4 | −00 56 | 250 | 1.1 | 5.6 + 7.5 | Binary, 217.2 years |

## Planetary nebula

| NGC/IC | Other | Con | RA h m | Dec ° ′ | Mag p | Diam ″ | Mag* |
|---|---|---|---|---|---|---|---|
| 246 | | Cet | 00 47.0 | −11 53 | 8.0 | 225 | 11.9 |

## Galaxies

| NGC/IC | Other | Con | RA h m | Dec ° ′ | Mag | Size ′ | Type |
|---|---|---|---|---|---|---|---|
| 7814 | | Peg | 00 03.3 | +16 09 | 10.5 | 6.3 × 2.6 | Sb |
| 128 | | Psc | 00 29.3 | +02 52 | 11.6 | 3.4 × 1.0 | S0 |
| 157 | | Cet | 00 34.8 | −08 24 | 10.4 | 4.3 × 2.9 | Sc |
| 210 | | Cet | 00 40.6 | −13 52 | 10.9 | 5.4 × 3.7 | Sb |
| 255 | | Cet | 00 47.8 | −11 28 | 11.8 | 3.1 × 2.8 | Sb |
| 309 | | Cet | 00 56.7 | −09 55 | 11.8 | 3.1 × 2.7 | Sc |
| 337 | | Cet | 00 59.8 | −07 35 | 11.6 | 2.8 × 2.0 | Sc |
| I.1613 | | Cet | 01 04.8 | +02 07 | 9.3 | 12.0 × 11.2 | Irr |
| 428 | | Cet | 01 12.9 | +00 59 | 11.4 | 4.1 × 3.2 | Sc |
| 474 | | Psc | 01 20.1 | +03 25 | 11.1 | 7.9 × 7.2 | S0 |
| 488 | | Psc | 01 21.8 | +05 15 | 10.3 | 5.2 × 4.1 | Sb |
| 524 | | Psc | 01 24.8 | +09 32 | 10.6 | 3.2 × 3.2 | E1 |
| 584 | | Cet | 01 31.1 | −06 52 | 10.4 | 3.8 × 2.4 | E4 |
| 628 | M74 | Psc | 01 36.7 | +15 47 | 9.2 | 10.2 × 9.5 | Sc |
| 676 | | Psc | 01 49.0 | +05 54 | 11.0 | 4.3 × 1.5 | Sa |
| 720 | | Cet | 01 53.0 | −13 44 | 10.2 | 4.4 × 2.8 | E3 |
| 772 | | Ari | 01 59.3 | +19 01 | 10.3 | 7.1 × 4.5 | Sb |
| 779 | | Cet | 01 57.7 | −05 58 | 11.1 | 4.1 × 4.5 | Sb |
| 864 | | Cet | 02 15.5 | +06 00 | 11.0 | 4.6 × 3.5 | Sc |
| 895 | | Cet | 02 21.6 | −05 31 | 11.8 | 3.6 × 2.8 | Sb |
| 936 | | Cet | 02 27.6 | −01 09 | 10.1 | 5.2 × 4.4 | SBa |
| 988 | | Cet | 02 35.4 | −09 21 | 11: | 4.7 × 2.7 | SBc |
| 1042 | | Cet | 02 40.4 | −08 26 | 10.9 | 4.7 × 3.9 | Sc |
| 1052 | | Cet | 02 41.1 | −08 15 | 10.6 | 2.9 × 2.0 | E2 |
| 1055 | | Cet | 02 41.8 | +00 26 | 10.6 | 7.6 × 3.0 | Sb |
| 1068 | M77 | Cet | 02 42.7 | −00 01 | 8.8 | 6.9 × 5.9 | Sb |
| 1073 | | Cet | 02 43.7 | +01 23 | 11.0 | 4.9 × 4.6 | SBc |
| 1084 | | Eri | 02 46.0 | −07 35 | 10.6 | 2.9 × 1.5 | Sc |
| 1087 | | Cet | 03 46.4 | −00 30 | 11.1 | 3.5 × 2.3 | Sc |
| 1179 | | Eri | 03 02.6 | −18 54 | 11.8 | 4.6 × 3.9 | S |
| 1300 | | Eri | 03 19.7 | −19 25 | 10.4 | 6.5 × 4.3 | SBb |
| 1337 | | Eri | 03 28.1 | −08 23 | 11.7 | 6.8 × 2.0 | S |
| 1407 | | Eri | 03 40.2 | −18 35 | 9.8 | 2.5 × 2.5 | E0 |

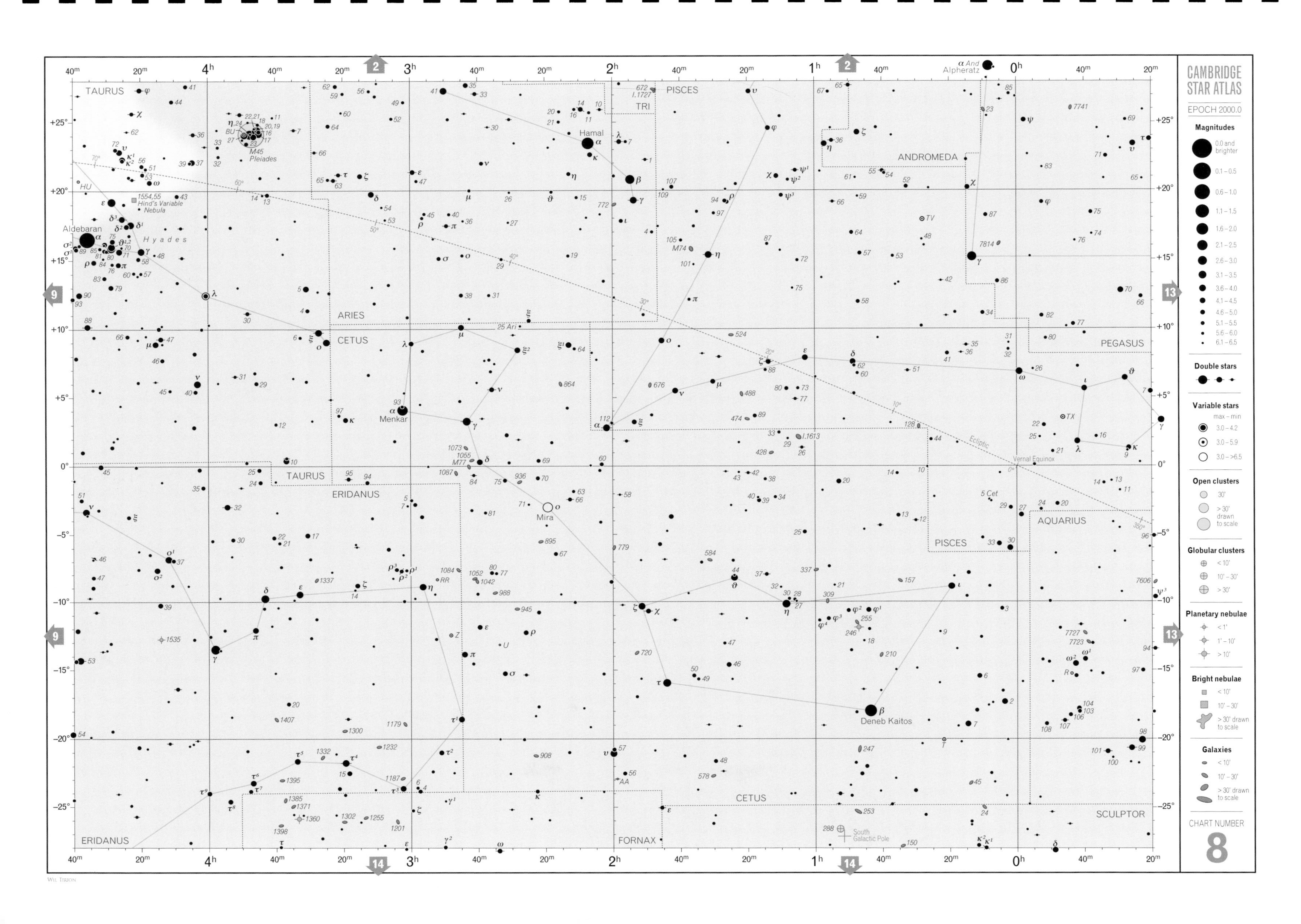

CAMBRIDGE STAR ATLAS
EPOCH 2000.0
Magnitudes
0.0 and brighter
0.1 – 0.5
0.6 – 1.0
1.1 – 1.5
1.6 – 2.0
2.1 – 2.5
2.6 – 3.0
3.1 – 3.5
3.6 – 4.0
4.1 – 4.5
4.6 – 5.0
5.1 – 5.5
5.6 – 6.0
6.1 – 6.5
Double stars
Variable stars
max – min
3.0 – 4.2
3.0 – 5.9
3.0 – >6.5
Open clusters
30'
> 30' drawn to scale
Globular clusters
< 10'
10' – 30'
> 30'
Planetary nebulae
< 1'
1' – 10'
> 10'
Bright nebulae
< 10'
10' – 30'
> 30' drawn to scale
Galaxies
< 10'
10' – 30'
> 30' drawn to scale
CHART NUMBER
8
TAURUS
ARIES
CETUS
PISCES
TRI
ANDROMEDA
PEGASUS
AQUARIUS
ERIDANUS
FORNAX
SCULPTOR
Aldebaran
Hyades
M45 Pleiades
1554,55 Hind's Variable Nebula
Hamal
Menkar
Mira
Deneb Kaitos
α And Alpheratz
Vernal Equinox
Ecliptic
South Galactic Pole
M74
M77
25 Ari
5 Cet
WIL TIRION

### Variable stars

| Star | Con | RA h m | Dec ° ′ | Range | Type | Period (days) | Spectrum | |
|---|---|---|---|---|---|---|---|---|
| λ | Tau | 04 00.7 | +12 29 | 3.3–3.8 | EA | 3.95 | B + A | |
| R | Lep | 04 59.6 | –14 18 | 5.5–11.7 | M | 432.1 | M | |
| W | Ori | 05 05.4 | +01 11 | 5.9–7.7 | SR | 212 | B | |
| RX | Lep | 05 11.4 | –11 51 | 5.0–7.0 | Irr | — | B | |
| CK | Ori | 05 30.3 | +04 12 | 5.9–7.1 | SR | 120: | M | |
| α | Ori | 05 55.2 | +07 24 | 0.4–1.3 | SR | 2110 | M | Betelgeuse |
| V | Mon | 06 22.7 | –02 12 | 6.0–13.7 | M | 333.8 | M | |
| T | Mon | 06 25.2 | +07 05 | 5.6–6.6 | Cep | 27.02 | F – K | |
| BL | Ori | 06 25.5 | +14 43 | 6.3–6.90 | Irr | — | M | |
| R | CMa | 07 19.5 | –16 24 | 5.7–6.3 | EA | 1.14 | F | |
| U | Mon | 07 30.8 | –09 47 | 5.9–7.8 | RV | 92.26 | F – K | |
| BN | Gem | 07 37.1 | +16 54 | 6.0–6.6 p | Irr | — | O | |

### Double stars

| Star | Con | RA h m | Dec ° ′ | PA ° | Sep ″ | Magnitudes | |
|---|---|---|---|---|---|---|---|
| 47 | Tau | 04 13.9 | +09 16 | 351 | 1.1 | 4.9 + 7.4 | |
| ϑ¹, ϑ² | Tau | 04 28.7 | +15 52 | 346 | 337.4 | 3.4 + 3.8 | |
| σ¹, σ² | Tau | 04 39.3 | +15 55 | 193 | 431.2 | 4.7 + 5.1 | |
| 55 | Eri | 04 43.6 | –08 48 | 317 | 9.2 | 6.7 + 6.8 | |
| 14 | Ori | 05 07.9 | +08 30 | 322 | 0.8 | 5.8 + 6.5 | Binary, 198.9 years |
| ρ | Ori | 05 13.3 | +02 52 | 64 | 7.0 | 4.5 + 8.3 | |
| κ | Lep | 05 13.2 | –12 56 | 358 | 2.6 | 4.5 + 7.4 | |
| β | Ori | 05 14.5 | –08 12 | 202 | 9.5 | 0.2 + 6.8 | Rigel |
| η | Ori | 05 24.5 | –02 24 | 80 | 1.5 | 3.8 + 4.5 | |
| ϑ¹ | Ori | 05 35.3 | –05 23 | 31 | 8.8 | 6.7 + 7.9 | Trapezium; in Orion Nebula |
| | | | | 132 | 12.8 | 5.1 | |
| | | | | 96 | 21.5 | 6.7 | |
| ϑ² | Ori | 05 35.4 | –05 25 | 92 | 52.5 | 5.2 + 6.5 | In Orion Nebula |
| ι | Ori | 05 35.4 | –05 55 | 141 | 11.3 | 2.8 + 6.9 | |
| ζ | Ori | 05 40.8 | –01 57 | 165 | 2.3 | 1.9 + 4.0 | Alnitak; binary, 1509 years |
| 75 | Ori | 06 17.1 | +09 57 | 258 | 62.7 | 5.4 + 9.5 | |
| | | | | 159 | 117.3 | 8.5 | |
| 8 | Mon | 06 23.8 | +04 36 | 27 | 13.4 | 4.5 + 6.5 | = ε Mon |
| β | Mon | 06 28.8 | –07 02 | 132 | 7.3 | 4.7 + 5.2 | |
| | | | | 124 | 10.0 | 6.1 | |
| ν¹ | CMa | 06 36.4 | –18 40 | 262 | 17.5 | 5.8 + 8.5 | |
| 15 | Mon | 06 41.0 | +09 54 | 213 | 2.8 | 4.7 + 7.5 | S Mon |
| | | | | 13 | 16.6 | 9.8 | |
| | | | | 308 | 41.3 | 9.6 | In cluster 2264 |
| | | | | 139 | 73.9 | 9.9 | |
| | | | | 222 | 156.0 | 7.7 | |
| α | CMa | 06 45.2 | –16 43 | 150 | 4.6 | –1.4 + 8.5 | Sirius; binary 50.1 years |
| 38 | Gem | 06 54.6 | +13 11 | 145 | 7.1 | 4.7 + 7.7 | Binary, 3,190 years |
| μ | CMa | 06 56.1 | –14 03 | 340 | 3.0 | 5.3 + 8.6 | |

### Open clusters

| NGC/IC | Other | Con | RA h m | Dec ° ′ | Mag | Diam ′ | N* | |
|---|---|---|---|---|---|---|---|---|
| 1647 | | Tau | 04 46.0 | +19 04 | 6.4 | 45 | 200 | |
| 1662 | | Ori | 04 48.5 | +10 56 | 6.4 | 20 | 35 | |
| 1807 | | Tau | 05 10.7 | +16 32 | 7.0 | 17 | 20 | |
| 1817 | | Tau | 05 12.1 | +16 42 | 7.7 | 16 | 60 | |
| 1981 | | Ori | 05 35.2 | –04 26 | 4.6 | 25 | 20 | |
| 2169 | | Ori | 06 08.4 | +13 57 | 5.9 | 7 | 30 | |
| 2194 | | Ori | 06 13.8 | +12 48 | 8.5 | 10 | 80 | |
| 2204 | | CMa | 06 15.7 | –18 39 | 8.6 | 13 | 80 | |
| 2215 | | Mon | 06 21.0 | –07 17 | 8.4 | 11 | 40 | |
| 2232 | | Mon | 06 26.6 | –04 45 | 3.9 | 30 | 20 | Contains 10 Mon |
| 2244 | | Mon | 06 32.4 | +04 52 | 4.8 | 24 | 100 | In Rosette Nebula |
| 2252 | | Mon | 06 35.0 | +05 23 | 7.7 | 20 | 30 | |
| 2286 | | Mon | 06 47.6 | –03 10 | 7.5 | 15 | 50 | Asterism? |
| 2301 | | Mon | 06 51.8 | +00 28 | 6.0 | 12 | 80 | |
| 2323 | M50 | Mon | 07 03.2 | –08 20 | 5.9 | 16 | 80 | |
| 2335 | | Mon | 07 06.6 | –10 05 | 7.2 | 12 | 35 | |
| 2343 | | Mon | 07 08.3 | –10 39 | 6.7 | 7 | 20 | |
| 2345 | | CMa | 07 08.3 | –13 10 | 7.7 | 12 | 70 | |
| 2353 | | Mon | 07 14.6 | –10 18 | 7.1 | 20 | 30 | |
| 2360 | | CMa | 07 17.8 | –15 37 | 7.2 | 13 | 80 | |
| 2374 | | CMa | 07 24.0 | –13 16 | 8.0 | 19 | 25 | |
| 2395 | | Gem | 07 27.1 | +13 35 | 8.0 | 12 | 30 | |
| 2414 | | Pup | 07 33.3 | –15 27 | 7.9 | 4 | 35 | |
| 2422 | M47 | Pup | 07 36.6 | –14 30 | 4.4 | 30 | 30 | |
| 2423 | | Pup | 07 37.1 | –13 52 | 6.7 | 19 | 40 | |
| — | Mel71 | Cep | 07 37.5 | –12 04 | 7.1 | 9 | 80 | |
| 2437 | M46 | Pup | 07 41.8 | –14 49 | 6.1 | 27 | 100 | Contains pl. neb. 2438 |

### Bright diffuse nebulae

| NGC/IC | Other | Con | RA h m | Dec ° ′ | Type | Diam ′ | Mag* | |
|---|---|---|---|---|---|---|---|---|
| 1554, 1555 | | Tau | 04 21.8 | +19 32 | R | Var | 9.4 | Hind's Variable Nebula |
| 1973 | | Ori | 05 35.1 | –04 44 | E + R | 5 × 5 | 7.4 | |
| 1975 | | Ori | 05 35.4 | –04 41 | E + R | 10 × 5 | 10.9 | |
| 1976 | M42 | Ori | 05 35.4 | –05 27 | E + R | 66 × 60 | 2.9 | Great Orion Nebula |
| 1977 | | Ori | 05 35.5 | –04 52 | E + R | 20 × 10 | 4.6 | |
| 1982 | M43 | Ori | 05 35.6 | –05 16 | E + R | 20 × 15 | 6.9 | Extension of M42 |
| I.434 | | Ori | 05 41.0 | –02 24 | E | 60 × 10 | 2.1 | Contains Horse Head Nebula (B33) |
| 2024 | | Ori | 05 40.7 | –02 27 | E | 30 × 30 | 2.1 | ζ Ori |
| 2068 | M78 | Ori | 05 46.7 | +00 03 | R | 8 × 6 | 10.5 | |
| 2237–2239 | | Mon | 06 32.3 | +05 03 | E | 80 × 60 | | Rosetta Nebula |
| 2261 | R Mon | Mon | 06 39.2 | +08 44 | E + R | 2 × 1 | 10.0 | Hubble's Variable Nebula |
| 2264 | S Mon | Cep | 06 40.9 | +09 54 | E | 60 × 30 | 4.5 | Includes Cone Nebula |
| I.2177 | | Mon | 07 05.1 | –10 42 | E | 120 × 40 | 6.2 | |

### Planetary nebulae

| NGC/IC | Other | Con | RA h m | Dec ° ′ | Mag p | Diam ″ | Mag* | |
|---|---|---|---|---|---|---|---|---|
| 1535 | | Eri | 04 14.2 | –12 44 | 9.6 | 18/44 | 12.2 | |
| I.418 | | Lep | 05 27.5 | –12 42 | 10.7 | 12 | 10.6 | |
| 2438 | | Pup | 07 41.8 | –14 44 | 10.1 | 66 | 17.7 | In open cluster 2437 |
| 2440 | | Pup | 07 41.9 | –18 13 | 10.8 | 14/32 | 14.3 | |

### Galaxies

| NGC/IC | Other | Con | RA h m | Dec ° ′ | Mag | Size ′ | Type |
|---|---|---|---|---|---|---|---|
| 1637 | | Eri | 04 41.5 | –02 51 | 10.9 | 3.3 × 2.9 | Sc |
| 1723 | | Eri | 04 59.4 | –10 59 | 12: | 3.7 × 2.3 | SB |

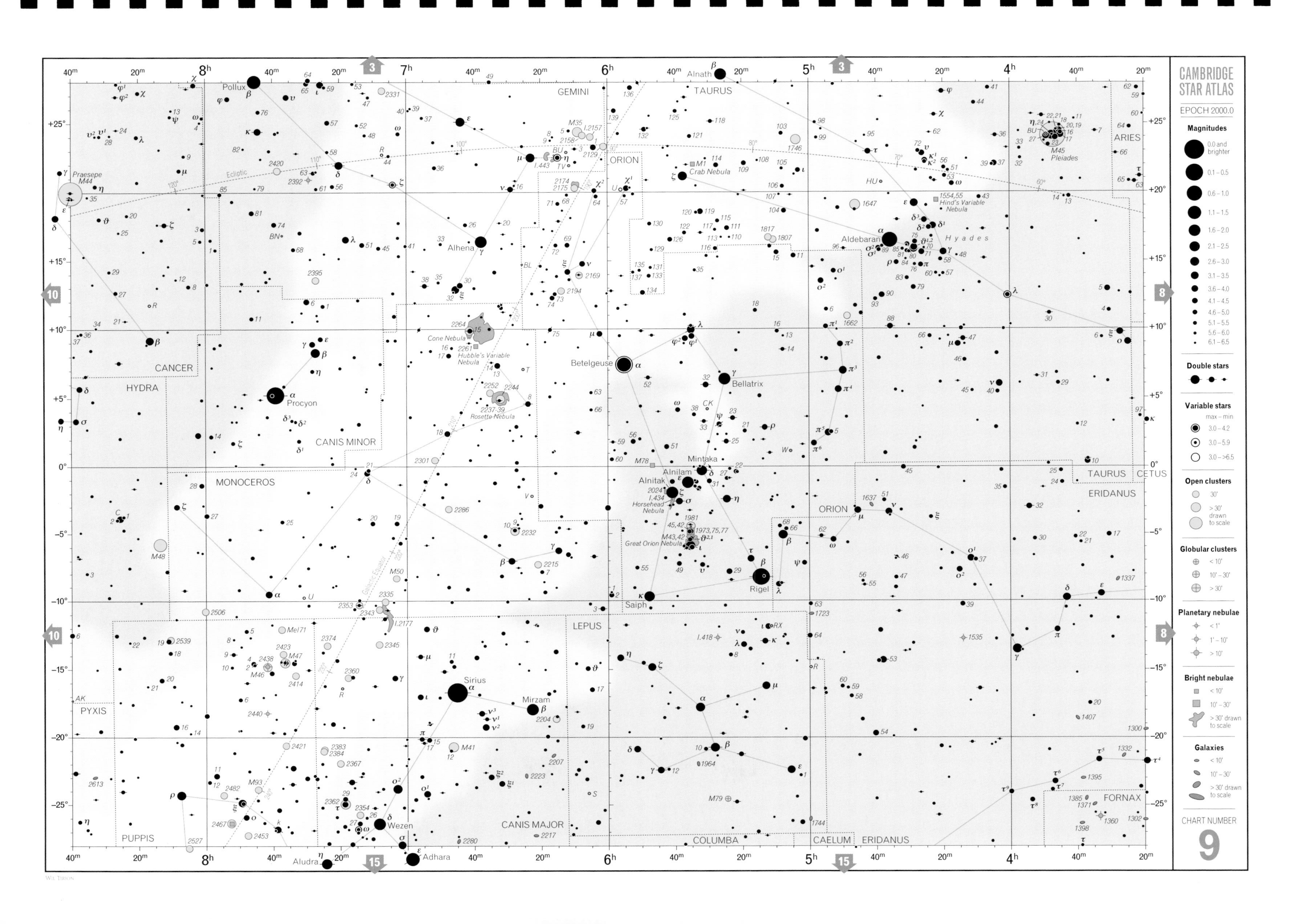

CAMBRIDGE STAR ATLAS
EPOCH 2000.0
Magnitudes
0.0 and brighter
0.1 – 0.5
0.6 – 1.0
1.1 – 1.5
1.6 – 2.0
2.1 – 2.5
2.6 – 3.0
3.1 – 3.5
3.6 – 4.0
4.1 – 4.5
4.6 – 5.0
5.1 – 5.5
5.6 – 6.0
6.1 – 6.5
Double stars
Variable stars
max – min
3.0 – 4.2
3.0 – 5.9
3.0 – >6.5
Open clusters
30'
> 30'
drawn to scale
Globular clusters
< 10'
10' – 30'
> 30'
Planetary nebulae
< 1'
1' – 10'
> 10'
Bright nebulae
< 10'
10' – 30'
> 30' drawn to scale
Galaxies
< 10'
10' – 30'
> 30' drawn to scale
CHART NUMBER
9
GEMINI
TAURUS
ARIES
ORION
CANCER
HYDRA
CANIS MINOR
MONOCEROS
PYXIS
PUPPIS
LEPUS
CANIS MAJOR
COLUMBA
CAELUM
ERIDANUS
TAURUS
CETUS
FORNAX
Pollux
Praesepe M44
Procyon
Alhena
Betelgeuse
Bellatrix
Aldebaran
Alnath
Hyades
M45 Pleiades
Crab Nebula
Hind's Variable Nebula
Cone Nebula
Hubble's Variable Nebula
Rosette Nebula
Mintaka
Alnilam
Alnitak
Horsehead Nebula
Great Orion Nebula
Rigel
Saiph
Sirius
Mirzam
Wezen
Adhara
Aludra
Ecliptic
Galactic Equator

## Chart 10 *RA 8$^h$ to 12$^h$, declination +20° to −20°*

### Variable stars

| Star | Con | RA h m | Dec ° ′ | Range | Type | Period (days) | Spectrum |
|---|---|---|---|---|---|---|---|
| R | Cnc | 08 16.6 | +11 44 | 6.1–11.8 | M | 361.6 | M |
| AK | Hya | 08 39.9 | −17 18 | 6.3–6.9 | SR | 112: | M |
| X | Cnc | 08 55.4 | +17 14 | 5.6–7.5 | SR | 195: | K |
| R | Leo | 09 47.6 | +11 26 | 4.4–11.3 | M | 312.4 | M |
| U | Hya | 10 37.6 | −13 23 | 4.3–6.5 | SR | 450: | B |

### Double stars

| Star | Con | RA h m | Dec ° ′ | PA ° | Sep ″ | Magnitudes | |
|---|---|---|---|---|---|---|---|
| ζ | Cnc | 08 12.2 | +17 39 | 72 | 6.0 | 5.6 + 6.2 | Binary, 1150 years |
| | | | | 86 | 0.8 | 6.0 | Binary, 59.7 years |
| ε | Hya | 08 46.8 | +06 25 | 302 | 2.7 | 3.8 + 4.7 | Binary, 890 years |
| 27 | Hya | 09 20.5 | −09 33 | 211 | 229.4 | 5.0 + 6.9 | |
| ω | Leo | 09 28.5 | +09 03 | 84 | 0.6 | 5.9 + 6.5 | Binary, 118.2 years |
| 6 | Leo | 09 32.0 | +09 43 | 75 | 37.4 | 5.2 + 8.2 | |
| γ | Sex | 09 52.5 | −08 06 | 56 | 0.6 | 5.6 + 6.1 | Binary, 75.6 years |
| α | Leo | 10 08.4 | +11 58 | 307 | 176.9 | 1.4 + 7.7 | Regulus |
| γ | Leo | 10 20.0 | +19 51 | 125 | 4.4 | 2.2 + 3.5 | Algieba; binary, 618.6 years |
| | | | | 291 | 259.9 | 9.2 | |
| | | | | 302 | 333.0 | 9.6 | |
| 49 | Leo | 10 35.0 | +08 39 | 157 | 2.4 | 5.8 + 8.5 | TX Leo |
| ι | Leo | 11 23.9 | +10 32 | 117 | 1.7 | 4.0 + 6.7 | Binary, 192 years |
| γ | Crt | 11 24.9 | −17 41 | 96 | 5.2 | 4.1 + 9.6 | |
| 90 | Leo | 11 34.7 | +16 48 | 209 | 3.3 | 6.0 + 7.3 | |
| | | | | 234 | 63.1 | 8.7 | |

### Open clusters

| NGC/IC | Other | Con | RA h m | Dec ° ′ | Mag | Diam ′ | N* | |
|---|---|---|---|---|---|---|---|---|
| 2506 | | Mon | 08 00.2 | −10 47 | 7.6 | 7 | 150 | |
| 2539 | | Pup | 08 10.7 | −12 50 | 6.5 | 22 | 50 | |
| 2548 | M48 | Hya | 08 13.8 | −05 48 | 5.8 | 54 | 80 | |
| 2632 | M44 | Cnc | 08 40.1 | +19 59 | 3.1 | 95 | 50 | Praesepe; Beehive |
| 2682 | M67 | Cnc | 08 50.4 | +11 49 | 6.9 | 30 | 200 | |

### Planetary nebula

| NGC/IC | Other | Con | RA h m | Dec ° ′ | Mag p | Diam ″ | Mag* | |
|---|---|---|---|---|---|---|---|---|
| 3242 | | Hya | 10 24.8 | −18 38 | 8.6 | 26/40 | 12.0 | Ghost of Jupiter |

### Galaxies

| NGC/IC | Other | Con | RA h m | Dec ° ′ | Mag | Size ′ | Type | |
|---|---|---|---|---|---|---|---|---|
| 2775 | | Cnc | 09 10.3 | +07 02 | 10.3 | 4.5 × 3.5 | Sa | |
| 2967 | | Sex | 09 42.1 | +00 20 | 11.6 | 3.0 × 2.9 | Sc | |
| 2974 | | Sex | 09 42.6 | −03 42 | 10.8 | 3.4 × 2.1 | Sa | |
| — | U5373 | Sex | 10 00.0 | +05 20 | 11.4 | 4.6 × 3.3 | Irr | |
| 3115 | | Sex | 10 05.2 | −07 43 | 9.2 | 8.3 × 3.2 | E6 | |
| — | U5470 | Leo | 10 08.4 | +12 18 | 9.8 | 10.7 × 8.3 | dE3 | Leo I |
| 3166 | | Sex | 10 13.8 | +03 26 | 10.6 | 5.2 × 2.7 | SBa | |
| 3169 | | Sex | 10 14.2 | +03 28 | 10.5 | 4.8 × 3.2 | Sb | |
| 3351 | M95 | Leo | 10 44.0 | +11 42 | 9.7 | 7.4 × 5.1 | SBb | |
| 3367 | | Leo | 10 46.6 | +13 45 | 11.5 | 2.3 × 2.1 | Sc | |
| 3368 | M96 | Leo | 10 46.8 | +11 49 | 9.2 | 7.1 × 5.1 | Sb | |
| 3377 | | Leo | 10 47.7 | +13 59 | 10.2 | 4.4 × 2.7 | E5 | |
| 3379 | M105 | Leo | 10 47.8 | +12 35 | 9.3 | 4.5 × 4.0 | E1 | |
| 3384 | | Leo | 10 48.3 | +12 38 | 10.0 | 5.9 × 2.6 | E7 | |
| 3412 | | Leo | 10 50.9 | +13 25 | 10.6 | 3.6 × 2.0 | E5 | |
| 3489 | | Leo | 11 00.3 | +13 54 | 10.3 | 3.7 × 2.1 | E6 | |
| 3507 | | Leo | 11 03.5 | +18 08 | 11.4 | 3.5 × 3.0 | SBb | |
| 3521 | | Leo | 11 05.8 | −00 02 | 8.9 | 9.5 × 5.0 | Sb | |
| 3593 | | Leo | 11 14.6 | +12 49 | 11.0 | 5.8 × 2.5 | Sb | |
| 3596 | | Leo | 11 15.1 | +14 47 | 11.6 | 4.2 × 4.1 | Sc | |
| 3607 | | Leo | 11 16.9 | +18 03 | 10.0 | 3.7 × 3.2 | E1 | |
| 3623 | M65 | Leo | 11 18.9 | +13 05 | 9.3 | 10.0 × 3.3 | Sb | |
| 3626 | | Leo | 11 20.1 | +18 21 | 10.9 | 3.1 × 2.2 | Sb | |
| 3627 | M66 | Leo | 11 20.2 | +12 59 | 9.0 | 8.7 × 4.4 | Sb | |
| 3628 | | Leo | 11 20.3 | +13 36 | 9.5 | 14.8 × 3.6 | Sb | |
| 3640 | | Leo | 11 21.1 | +03 14 | 10.3 | 4.1 × 3.4 | E1 | |
| 3655 | | Leo | 11 22.9 | +16 35 | 11.6 | 1.6 × 1.1 | S: | |
| 3672 | | Crt | 11 25.0 | −09 48 | 11.5 | 4.1 × 2.1 | Sb | |
| 3686 | | Leo | 11 27.7 | +17 13 | 11.4 | 3.3 × 2.6 | Sc | |
| 3810 | | Leo | 11 41.0 | +11 28 | 10.8 | 4.3 × 3.1 | Sc | |
| 3818 | | Vir | 11 42.0 | −06 09 | 11.8 | 2.1 × 1.4 | E2 | |
| 3887 | | Crt | 11 47.1 | −16 51 | 11.0 | 3.3 × 2.7 | Sc | |
| 3962 | | Crt | 11 54.7 | −13 58 | 10.6 | 2.9 × 2.6 | E2 | |
| 4027 | | Crv | 11 59.5 | −19 16 | 11.1 | 3.0 × 2.3 | Sc | |

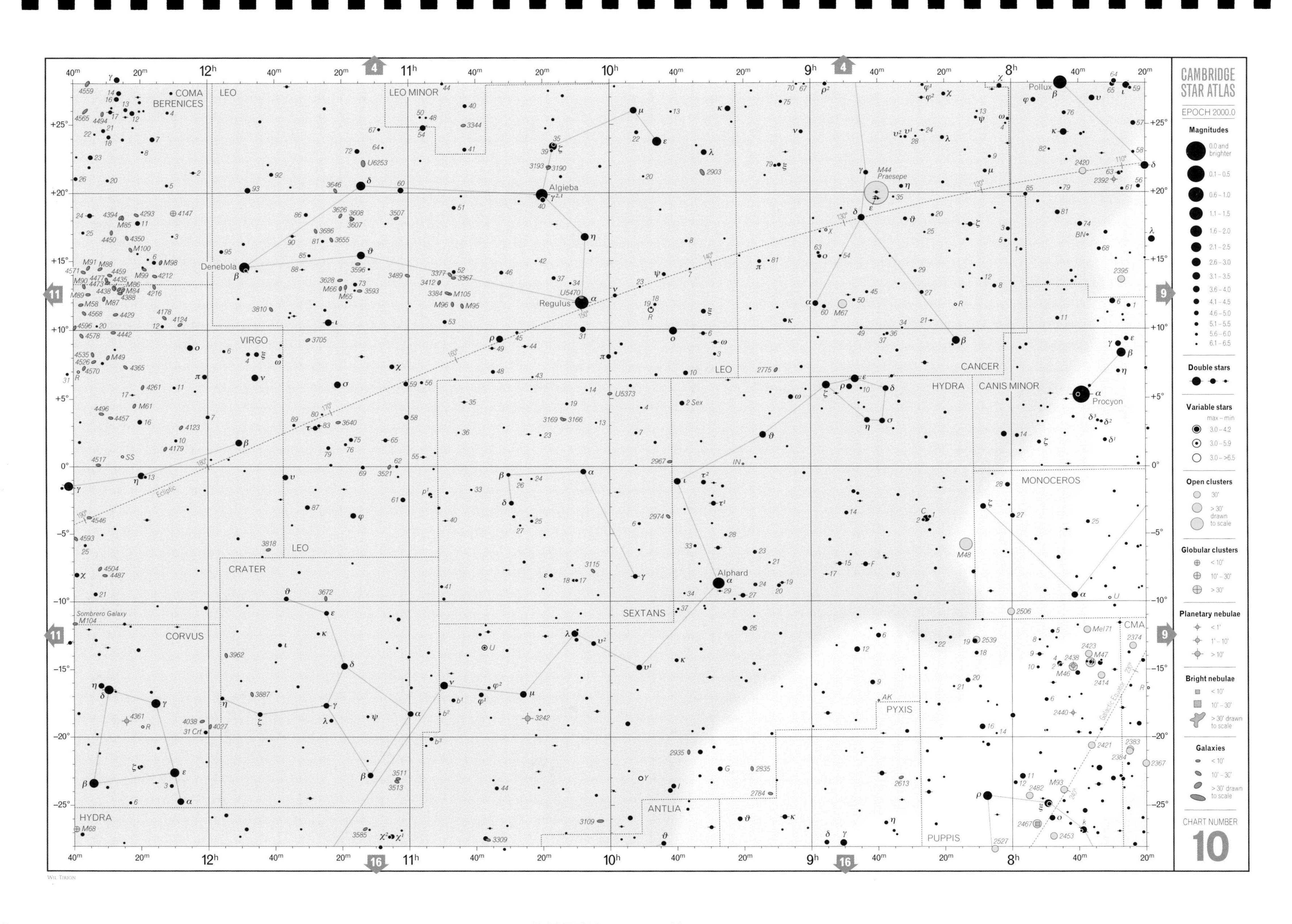

CAMBRIDGE STAR ATLAS
EPOCH 2000.0
Magnitudes
0.0 and brighter
0.1 – 0.5
0.6 – 1.0
1.1 – 1.5
1.6 – 2.0
2.1 – 2.5
2.6 – 3.0
3.1 – 3.5
3.6 – 4.0
4.1 – 4.5
4.6 – 5.0
5.1 – 5.5
5.6 – 6.0
6.1 – 6.5
Double stars
Variable stars
max – min
3.0 – 4.2
3.0 – 5.9
3.0 – >6.5
Open clusters
30'
> 30'
drawn to scale
Globular clusters
< 10'
10' – 30'
> 30'
Planetary nebulae
< 1'
1' – 10'
> 10'
Bright nebulae
< 10'
10' – 30'
> 30' drawn to scale
Galaxies
< 10'
10' – 30'
> 30' drawn to scale
CHART NUMBER
10
COMA BERENICES
LEO
LEO MINOR
VIRGO
CANCER
HYDRA
CANIS MINOR
MONOCEROS
SEXTANS
CRATER
CORVUS
ANTLIA
PYXIS
PUPPIS
CMA
Pollux
Procyon
Algieba
Regulus
Denebola
Alphard
M44 Praesepe
Sombrero Galaxy M104
Ecliptic
Galactic Equator
Wil Tirion

# Chart 11 *RA 12ʰ to 16ʰ, declination +20° to −20°*

## Variable stars

| Star | Con | RA h m | Dec ° ′ | Range | Type | Period (days) | Spectrum |
|---|---|---|---|---|---|---|---|
| R | Crv | 12 19.6 | −19 15 | 6.7–14.4 | M | 317.0 | M |
| SS | Vir | 12 25.3 | +00 48 | 6.0–9.6 | M | 354.7 | K |
| R | Vir | 12 38.50 | +06 59 | 6.0–12.1 | M | 145.6 | M |
| S | Vir | 13 33.0 | −07 12 | 6.3–13.2 | M | 377.4 | M |
| δ | Lib | 15 01.0 | −08 31 | 4.9–5.9 | EA | 2.33 | B |
| R | Ser | 15 50.7 | +15 08 | 5.2–14.4 | M | 356.4 | M |

## Double stars

| Star | Con | RA h m | Dec ° ′ | PA ° | Sep ″ | Magnitudes | |
|---|---|---|---|---|---|---|---|
| 17 | Vir | 12 22.5 | +05 18 | 337 | 20.0 | 6.6 + 9.4 | |
| δ | Crv | 12 29.9 | −16 31 | 214 | 24.2 | 3.0 + 9.2 | Algorab |
| 24 | Com | 12 35.1 | +18 23 | 271 | 20.3 | 5.2 + 6.7 | |
| γ | Vir | 12 41.7 | −01 27 | 267 | 1.8 | 3.5 + 3.5 | Porrima; binary, 171.4 years |
| θ | Vir | 13 09.9 | −05 32 | 343 | 7.1 | 4.4 + 9.4 | |
| 84 | Vir | 13 43.1 | +03 32 | 229 | 2.9 | 5.5 + 7.9 | |
| π | Boo | 14 40.7 | +16 25 | 108 | 5.6 | 4.9 + 5.8 | |
| ζ | Boo | 14 41.1 | +13 44 | 300 | 0.8 | 4.5 + 4.6 | Binary, 123.4 years |
| μ | Lib | 14 49.3 | −14 09 | 355 | 1.8 | 5.8 + 6.7 | |
| α², α¹ | Lib | 14 50.9 | −16 02 | 314 | 231.0 | 2.8 + 5.2 | Zubenalgenubi |
| ι | Lib | 15 12.2 | −19 47 | 111 | 57.8 | 5.1 + 9.4 | Binary, 22.4 years |
| δ | Ser | 15 34.8 | +10 32 | 176 | 4.4 | 4.2 + 5.2 | Binary, 3,168 years |
| 47 | Lib | 15 55.0 | −19 23 | 129 | 0.5 | 6.1 + 8.1 | |

## Globular clusters

| NGC/IC | Other | Con | RA h m | Dec ° ′ | Mag | Diam ′ |
|---|---|---|---|---|---|---|
| 4147 | | Com | 12 10.1 | +18 33 | 10.3 | 4.0 |
| 5024 | M53 | Com | 13 12.9 | +18 10 | 7.7 | 12.6 |
| 5053 | | Com | 13 16.4 | +17 42 | 9.8 | 10.5 |
| 5634 | | Vir | 14 29.6 | −05 59 | 9.6 | 4.9 |
| 5904 | M5 | Ser | 15 18.6 | +02 05 | 5.8 | 17.4 |

## Planetary nebula

| NGC/IC | Other | Con | RA h m | Dec ° ′ | Mag p | Diam ″ | Mag* |
|---|---|---|---|---|---|---|---|
| 4361 | | Crv | 12 24.5 | −18 48 | 10.3 | 45/110 | 13.2 |

## Galaxies

| NGC/IC | Other | Con | RA h m | Dec ° ′ | Mag | Size ′ | Type | |
|---|---|---|---|---|---|---|---|---|
| 4038 | | Crv | 12 01.9 | −18 52 | 10.7 | 2.6 × 1.8 | Sc | |
| 4123 | | Vir | 12 08.2 | +02 53 | 11.2 | 4.50 × 3.5 | SBb | |
| 4178 | | Vir | 12 12.8 | +10 52 | 11.4 | 5.0 × 2.0 | SBc | |
| 4179 | | Vir | 12 12.9 | +01 18 | 10.9 | 4.2 × 1.2 | S0 | |
| 4192 | M98 | Com | 12 13.8 | +14 54 | 10.1 | 9.5 × 3.2 | Sb | |
| 4212 | | Com | 12 15.7 | +13 54 | 11.2 | 3.0 × 2.1 | Sb | |
| 4216 | | Vir | 12 15.9 | +13 09 | 10.0 | 8.3 × 2.2 | Sb | |
| 4254 | M99 | Com | 12 18.8 | +14 25 | 9.8 | 5.4 × 4.8 | Sc | |
| 4261 | | Vir | 12 19.4 | +05 49 | 10.3 | 3.9 × 3.2 | E2 | |
| 4293 | | Com | 12 21.2 | +18 23 | 11.2 | 6.0 × 3.0 | Sa | |
| 4303 | M61 | Vir | 12 21.9 | +04 28 | 9.7 | 6.0 × 5.5 | Sc | |
| 4321 | M100 | Com | 12 22.9 | +15 49 | 9.4 | 6.9 × 6.2 | Sc | |
| 4350 | | Vir | 12 24.0 | +16 42 | 11.1 | 3.2 × 1.1 | E7 | |
| 4365 | | Vir | 12 24.5 | +07 19 | 10.5 | 6.2 × 4.6 | E2 | |
| 4374 | M84 | Vir | 12 25.1 | +12 53 | 9.3 | 5.0 × 4.4 | E1 | |
| 4382 | M85 | Com | 12 25.4 | +18 11 | 9.2 | 7.1 × 5.2 | E | |
| 4388 | | Vir | 12 25.8 | +12 40 | 11.1 | 5.1 × 1.4 | Sb | |
| 4394 | | Com | 12 25.9 | +18 13 | 10.9 | 3.9 × 3.5 | SBb | |
| 4406 | M86 | Vir | 12 26.2 | +12 57 | 9.2 | 7.4 × 5.5 | E3 | |
| 4429 | | Vir | 12 27.4 | +11 07 | 10.2 | 5.5 × 2.6 | S0 | |
| 4435 | | Vir | 12 27.7 | +13 05 | 10.9 | 3.0 × 1.0 | E4 | |
| 4438 | | Vir | 12 27.8 | +13 01 | 10.1 | 9.3 × 3.9 | Sa | |
| 4442 | | Vir | 12 28.1 | +09 48 | 10.5 | 4.6 × 1.9 | E5 | |
| 4450 | | Com | 12 28.5 | +17 05 | 10.1 | 4.8 × 3.5 | Sb | |
| 4457 | | Vir | 12 29.0 | +03 33 | 10.8 | 3.0 × 2.5 | SBa | |
| 4459 | | Com | 12 29.0 | +13 59 | 10.4 | 3.8 × 2.8 | E2 | |
| 4472 | M49 | Vir | 12 29.8 | +80 00 | 8.4 | 8.9 × 7.4 | E4 | |
| 4473 | | Vir | 12 29.8 | +13 26 | 10.2 | 4.5 × 2.6 | E4 | |
| 4477 | | Com | 12 30.0 | +13 38 | 10.4 | 4.0 × 3.5 | SBa | |
| 4486 | M87 | Vir | 12 30.8 | +12 24 | 8.6 | 7.2 × 6.8 | E1 | Virgo A; 3C274 |
| 4487 | | Vir | 12 31.1 | −08 03 | 11.5 | 4.1 × 3.0 | Sc | |
| 4496 | | Vir | 12 31.7 | +03 56 | 11.7 | 3.9 × 3.1 | SBc | |
| 4501 | M88 | Com | 12 32.0 | +14 25 | 9.5 | 6.9 × 3.9 | Sb | |
| 4504 | | Vir | 12 32.3 | −07 34 | 11.7 | 4.0 × 2.0 | Sc | |
| 4517 | | Vir | 12 32.8 | +00 07 | 10.5 | 10.2 × 1.9 | Sc | |
| 4526 | | Vir | 12 34.0 | +07 42 | 9.6 | 7.2 × 2.3 | E7 | |
| 4535 | | Vir | 12 34.3 | +08 12 | 9.8 | 6.8 × 5.0 | SBc | |
| 4548 | M91 | Com | 12 35.4 | +14 30 | 10.2 | 5.4 × 4.4 | SBb | |
| 4546 | | Vir | 12 35.5 | −03 48 | 10.3 | 3.5 × 1.7 | E6 | |
| 4552 | M89 | Vir | 12 35.7 | +12 33 | 9.8 | 4.2 × 4.2 | E0 | |
| 4568 | | Vir | 12 36.6 | +11 14 | 10.8 | 4.6 × 2.1 | Sc | |
| 4569 | M90 | Vir | 12 36.8 | +13 10 | 9.5 | 9.5 × 4.7 | Sb | |
| 4570 | | Vir | 12 36.9 | +07 15 | 10.9 | 4.1 × 1.3 | S0 | |
| 4571 | | Vir | 12 36.9 | +14 13 | 11.3 | 3.8 × 3.4 | Sc | |
| 4578 | | Vir | 12 37.5 | +09 33 | 11.4 | 3.6 × 2.8 | Sa | |
| 4579 | M58 | Vir | 12 37.7 | +11 49 | 9.8 | 5.4 × 4.4 | Sb | |
| 4596 | | Vir | 12 39.9 | +10 11 | 10.5 | 3.9 × 2.8 | SBa | |
| 4594 | M104 | Vir | 12 40.0 | −11 37 | 8.3 | 8.9 × 4.1 | Sb | Sombrero Galaxy |
| 4621 | M59 | Vir | 12 42.0 | +11 39 | 9.8 | 5.1 × 3.4 | E3 | |
| 4636 | | Vir | 12 42.8 | +02 41 | 9.6 | 6.2 × 5.0 | E1 | |
| 4643 | | Vir | 12 43.3 | +01 59 | 10.6 | 3.4 × 2.7 | SBa | |
| 4649 | M60 | Vir | 12 43.7 | +11 33 | 8.8 | 7.2 × 6.2 | E1 | |
| 4651 | | Com | 12 43.7 | +16 24 | 10.7 | 3.8 × 2.7 | Sc | |
| 4654 | | Vir | 12 44.0 | +13 08 | 10.5 | 4.7 × 3.0 | Sc | |
| 4665 | | Vir | 12 45.1 | +03 03 | 11.6 | 4.2 × 3.5 | SBa | |
| 4666 | | Vir | 12 45.1 | −00 28 | 10.8 | 4.5 × 1.5 | Sc | |
| 4689 | | Com | 12 47.8 | +13 46 | 10.9 | 4.0 × 3.5 | Sb | |
| 4691 | | Vir | 12 48.2 | −03 20 | 11.2 | 3.2 × 2.7 | SBb | |
| 4697 | | Vir | 12 48.6 | −05 48 | 9.3 | 6.0 × 3.8 | E4 | |
| 4698 | | Vir | 12 48.4 | +08 29 | 10.7 | 4.3 × 2.5 | Sb | |
| 4699 | | Vir | 12 49.0 | −08 40 | 9.6 | 3.5 × 2.7 | Sa | |
| 4710 | | Vir | 12 49.6 | +15 10 | 11.0 | 5.1 × 1.4 | S0 | |
| 4731 | | Vir | 12 51.0 | −06 24 | 11.3 | 6.5 × 3.4 | SBc | |
| 4754 | | Vir | 12 52.3 | +11 19 | 10.6 | 4.7 × 2.6 | SB0 | |
| 4753 | | Vir | 12 52.4 | −01 12 | 9.9 | 5.4 × 2.9 | Pec | |
| 4762 | | Vir | 12 52.9 | +11 14 | 10.2 | 8.7 × 1.6 | SB0 | |
| 4781 | | Vir | 12 54.4 | −10 32 | 11.8 | 3.5 × 1.8 | Sc | |
| 4818 | | Vir | 12 56.8 | −08 31 | 11.7 | 4.5 × 1.7 | SB | |
| 4856 | | Vir | 12 59.3 | −15 02 | 10.4 | 4.6 × 1.6 | SBa | |
| 4866 | | Vir | 12 59.5 | +14 10 | 11.0 | 6.5 × 1.5 | Sb | |
| 4900 | | Vir | 13 00.6 | +02 30 | 11.5 | 2.3 × 2.2 | Sc | |
| 4902 | | Vir | 13 01.0 | −14 31 | 11.2 | 3.0 × 2.8 | Sb | |
| 4939 | | Vir | 13 04.2 | −10 20 | 11.4 | 5.8 × 3.2 | Sb | |
| 4941 | | Vir | 13 04.2 | −05 33 | 11.1 | 3.7 × 2.1 | Sb | |
| 4958 | | Vir | 13 05.8 | −08 01 | 10.5 | 4.1 × 1.4 | E6 | |
| 4984 | | Vir | 13 09.0 | −15 31 | 11.8 | 2.8 × 2.2 | Sc | |
| 4995 | | Vir | 13 09.7 | −07 50 | 11.0 | 2.5 × 1.7 | Sb | |
| 5018 | | Vir | 13 13.0 | −19 31 | 10.8 | 2.6 × 2.1 | Sa | |
| 5044 | | Vir | 13 15.4 | −16 23 | 11.0 | 2.6 × 2.6 | E0 | |
| 5054 | | Vir | 13 17.0 | −16 38 | 11.3 | 5.0 × 3.1 | Sb | |
| 5170 | | Vir | 13 29.8 | −17 58 | 11.8 | 8.1 × 1.3 | Sb | |
| 5247 | | Vir | 13 38.1 | −17 53 | 10.5 | 5.4 × 4.7 | Sb | |
| 5248 | | Boo | 13 37.5 | +08 53 | 10.2 | 6.5 × 4.9 | Sc | |
| 5363 | | Vir | 13 56.1 | +05 15 | 10.2 | 4.2 × 2.7 | E | |
| 5364 | | Vir | 13 56.2 | +05 01 | 10.4 | 7.1 × 5.0 | Sb | |
| 5427 | | Vir | 14 03.4 | −06 02 | 11.4 | 2.5 × 2.3 | Sc | |
| 5566 | | Vir | 14 20.3 | +03 56 | 10.5 | 6.5 × 2.4 | Sb | |
| 5576 | | Vir | 14 21.1 | +03 16 | 10.9 | 3.2 × 2.2 | E2 | |
| 5668 | | Vir | 14 33.4 | +04 27 | 11.5 | 3.3 × 3.1 | Sc | |
| 5701 | | Vir | 14 39.2 | +05 22 | 11.8 | 4.7 × 4.5 | SB | |
| 5713 | | Vir | 14 40.2 | −00 17 | 11.4 | 2.8 × 2.5 | Sc | |
| 5746 | | Vir | 14 44.9 | +01 57 | 10.6 | 7.9 × 1.7 | Sb | |
| 5813 | | Vir | 15 01.2 | +01 42 | 10.7 | 3.6 × 2.8 | E1 | |
| 5838 | | Vir | 15 05.4 | +02 06 | 10.8 | 4.2 × 1.6 | Sa | |
| 5846 | | Vir | 15 06.8 | +01 36 | 10.2 | 3.4 × 3.2 | E0 | |
| 5885 | | Vir | 15 15.1 | −10 05 | 11.7 | 3.5 × 3.2 | Sc | |

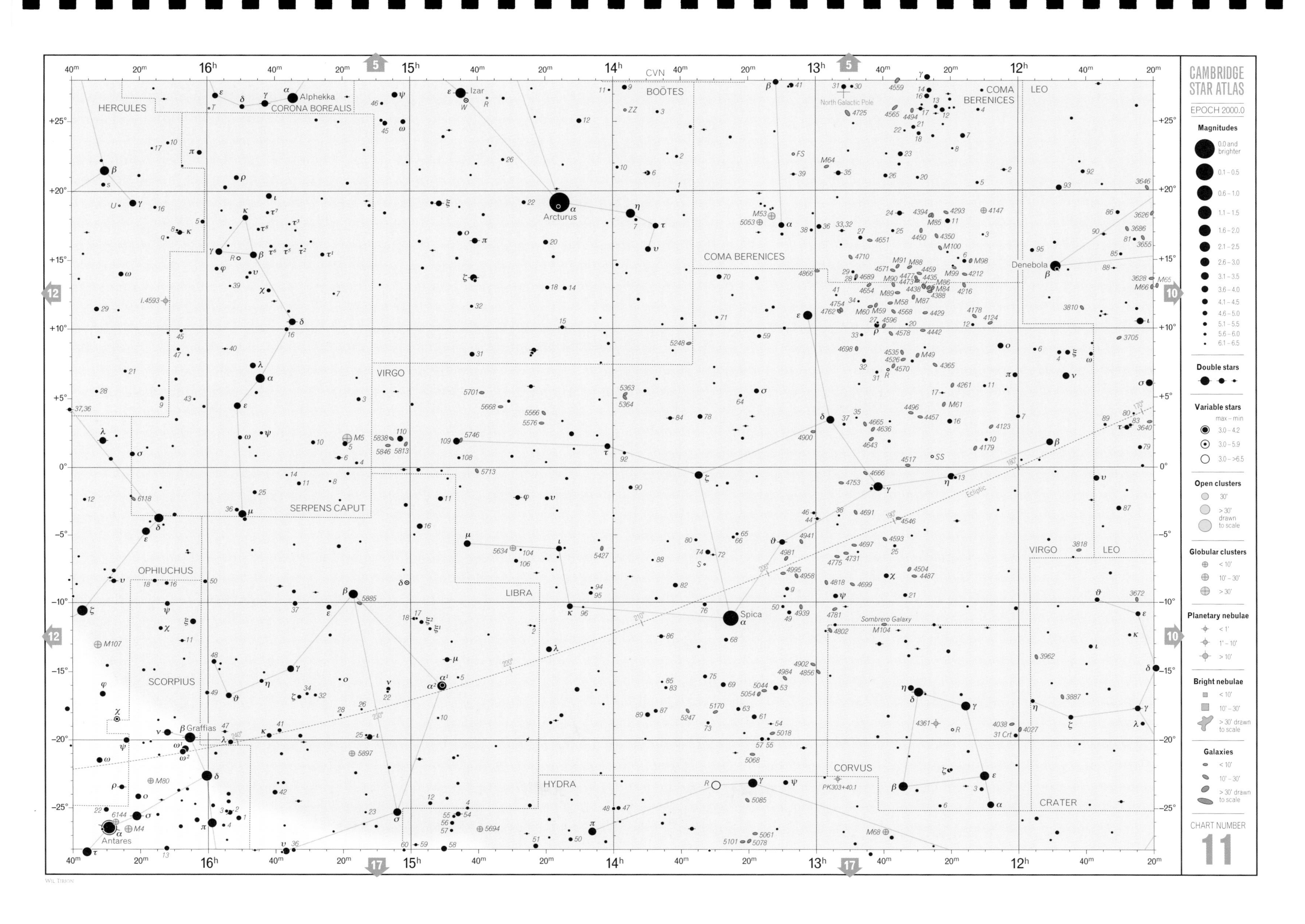

CAMBRIDGE STAR ATLAS
EPOCH 2000.0
Magnitudes
0.0 and brighter
0.1 – 0.5
0.6 – 1.0
1.1 – 1.5
1.6 – 2.0
2.1 – 2.5
2.6 – 3.0
3.1 – 3.5
3.6 – 4.0
4.1 – 4.5
4.6 – 5.0
5.1 – 5.5
5.6 – 6.0
6.1 – 6.5
Double stars
Variable stars
max – min
3.0 – 4.2
3.0 – 5.9
3.0 – >6.5
Open clusters
30'
> 30' drawn to scale
Globular clusters
< 10'
10' – 30'
> 30'
Planetary nebulae
< 1'
1' – 10'
> 10'
Bright nebulae
< 10'
10' – 30'
> 30' drawn to scale
Galaxies
< 10'
10' – 30'
> 30' drawn to scale
CHART NUMBER
11
HERCULES
CORONA BOREALIS
BOÖTES
CVN
COMA BERENICES
LEO
VIRGO
SERPENS CAPUT
OPHIUCHUS
LIBRA
SCORPIUS
HYDRA
CORVUS
CRATER
Alphekka
Izar
Arcturus
Denebola
Spica
Graffias
Antares
North Galactic Pole
Sombrero Galaxy
Ecliptic

# **Chart 12** *RA 16ʰ to 20ʰ, declination +20° to –20°*

## Variable stars

| Star | Con | RA | Dec | Range | Type | Period | Spectrum |
|---|---|---|---|---|---|---|---|
| | | h m | ° ′ | | | (days) | |
| U | Her | 16 25.8 | +18 54 | 6.5–13.4 | M | 406.1 | M |
| χ | Oph | 16 27.0 | –18 27 | 4.2–5.0 | Irr | — | M |
| V1010 | Oph | 16 49.5 | –15 40 | 6.1–7.0 | EB | 0.66 | A |
| S | Her | 16 51.9 | +14 56 | 6.4–13.8 | M | 307.4 | M |
| R | Oph | 17 17.8 | –16 06 | 7.0–13.8 | M | 302.6 | M |
| α | Her | 17 14.6 | +14 23 | 2.7–4.0 | SR | — | M |
| U | Oph | 17 16.5 | +01 13 | 5.9–6.6 | EA | 1.68 | B |
| Y | Sgr | 18 21.4 | –18 52 | 5.4–6.1 | Cep | 5.77 | F |
| 59 | Ser | 18 27.2 | +00 12 | 4.9–5.9 | ? | — | G + A |
| U | Sgr | 18 31.9 | –19 07 | 6.3–7.1 | Cep | 6.74 | F – G |
| X | Oph | 18 38.3 | +08 50 | 5.9–9.2 | M | 334.4 | M + K |
| V3879 | Sgr | 18 42.9 | –19 17 | 6.1–6.6 | SR | 50: | M |
| R | Sct | 18 47.5 | –05 42 | 4.5–8.2 | RV | 140.1 | G – K |
| FF | Aql | 18 18.2 | +17 22 | 5.2–5.7 | Cep | 4.47 | F |
| V | Aql | 19 04.4 | –05 41 | 6.6–8.4 | SR | 353 | K |
| R | Aql | 19 06.4 | +08 14 | 5.5–12.0 | M | 284.2 | M |
| U | Sge | 19 18.8 | +19 37 | 6.6–9.2 | EA | 3.38 | B + K |
| U | Aql | 19 29.4 | –07 03 | 6.1–6.9 | Cep | 7.02 | F – G |
| η | Aql | 19 52.5 | +01 00 | 3.5–4.4 | Cep | 7.17 | F – G |
| V505 | Sgr | 19 53.1 | –14 36 | 6.5–7.5 | EA | 1.18 | A + F |
| S | Sge | 19 56.0 | +16 38 | 5.3–6.0 | Cep | 8.38 | F – G |

## Double stars

| Star | Con | RA | Dec | PA | Sep | Magnitudes | |
|---|---|---|---|---|---|---|---|
| | | h m | ° ′ | ° | ″ | | |
| β | Sco | 16 05.4 | –19 48 | 21 | 13.6 | 2.6 + 4.9 | Graffias |
| | | | | 132 | 0.5 | 10.3 | |
| 11 | Sco | 16 07.6 | –12 45 | 257 | 3.3 | 5.6 + 9.9 | |
| κ | Her | 16 08.1 | –17 03 | 12 | 28.4 | 5.3 + 6.5 | |
| ν | Sco | 16 12.0 | –19 28 | 337 | 41.1 | 4.3 + 6.8 | |
| | | | | 3 | 0.9 | 6.8 | |
| | | | | 51 | 2.3 | 7.8 | |
| υ | Oph | 16 27.8 | –08 22 | 95 | 1.0 | 4.6 + 7.8 | |
| λ | Oph | 16 30.9 | +01 59 | 30 | 1.5 | 4.2 + 5.2 | Binary, 129.9 years |
| 37, 36 | Her | 16 40.6 | +04 13 | 230 | 69.8 | 5.8 + 7.0 | |
| 19 | Oph | 16 47.2 | +02 04 | 89 | 23.4 | 6.1 + 9.4 | |
| η | Oph | 17 10.4 | –15 43 | 237 | 0.6 | 3.0 + 3.5 | Binary, 84.3 years |
| α | Her | 17 14.6 | +14 23 | 104 | 4.6 | 2.9 v + 5.4 | Rasalgethi; var.; bin., 3,600 y. |
| 41 | Oph | 17 16.6 | –00 27 | 346 | 1.0 | 4.8 + 7.8 | |
| ν | Ser | 17 20.8 | –12 51 | 28 | 46.3 | 4.3 + 8.3 | |
| 53 | Oph | 17 34.6 | +09 35 | 191 | 41.2 | 5.8 + 8.5 | |
| τ | Oph | 18 03.1 | –08 11 | 283 | 1.7 | 5.2 + 5.9 | Binary, 280 years |
| | | | | 127 | 100.3 | 9.3 | |
| 70 | Oph | 18 05.5 | +02 30 | 148 | 3.8 | 4.2 + 6.0 | Binary, 88.1 years |
| 59 | Ser | 18 27.2 | +00 12 | 318 | 3.8 | 5.3 + 7.6 | |
| ϑ | Ser | 18 56.2 | +04 12 | 104 | 22.3 | 4.5 + 5.4 | |
| 23 | Aql | 19 18.5 | +01 05 | 5 | 3.1 | 5.3 + 9.3 | |
| π | Aql | 19 48.7 | +11 49 | 110 | 1.4 | 6.1 + 6.9 | |
| 57 | Aql | 19 54.6 | –08 14 | 170 | 35.7 | 5.8 + 6.5 | |

## Open clusters

| NGC/IC | Other | Con | RA | Dec | Mag | Diam | N* | |
|---|---|---|---|---|---|---|---|---|
| | | | h m | ° ′ | | ′ | | |
| I.4665 | | Oph | 17 46.3 | +05 43 | 4.2 | 41 | 30 | |
| 6494 | M23 | Sgr | 17 56.8 | –19 01 | 5.5 | 27 | 150 | |
| 6595 | | Sgr | 18 17.0 | –19 53 | 7.0 | 11 | 30 | |
| — | M24 | Sgr | 18 16.9 | –18 29 | 4.5 | 90 | — | Star cloud |
| 6604 | | Ser | 18 18.1 | –12 14 | 6.5 | 2 | 30 | |
| 6613 | M18 | Sgr | 18 19.9 | –17 08 | 6.9 | 9 | 20 | |
| 6633 | | Oph | 18 27.7 | +06 34 | 4.6 | 27 | 30 | |

## Open clusters *(continued)*

| NGC/IC | Other | Con | RA | Dec | Mag | Diam | N* | |
|---|---|---|---|---|---|---|---|---|
| | | | h m | ° ′ | | ′ | | |
| I.4725 | M25 | Sgr | 18 31.6 | –19 15 | 4.6 | 32 | 30 | |
| 6645 | | Sgr | 18 32.6 | –16 54 | 8.5 | 10 | 40 | |
| I.4756 | | Ser | 18 39.0 | +05 27 | 5.4 | 52 | 80 | |
| 6694 | M26 | Sct | 18 45.2 | –09 24 | 8.0 | 15 | 30 | |
| 6705 | M11 | Sct | 18 51.1 | –06 16 | 5.8 | 14 | 500 | Wild Duck Cluster |
| 6709 | | Aql | 18 51.5 | +10 21 | 6.7 | 13 | 40 | |
| 6716 | | Sgr | 18 54.6 | –19 53 | 6.9 | 7 | 20 | |
| 6738 | | Aql | 19 01.4 | +11 36 | 8.3 | 15 | — | |
| 6755 | | Aql | 19 07.8 | +04 14 | 7.5 | 15 | 100 | |
| — | H 20 | Sge | 19 53.1 | +18 20 | 7.7 | 7 | 15 | |

## Globular clusters

| NGC/IC | Other | Con | RA | Dec | Mag | Diam |
|---|---|---|---|---|---|---|
| | | | h m | ° ′ | | ′ |
| 6171 | M107 | Oph | 16 32.5 | –13 03 | 8.1 | 10.0 |
| 6218 | M12 | Oph | 16 47.2 | –01 57 | 6.6 | 14.5 |
| 6254 | M10 | Oph | 16 57.1 | –04 06 | 6.6 | 15.1 |
| 6333 | M9 | Oph | 17 19.2 | –18 31 | 7.9 | 9.3 |
| 6342 | | Oph | 17 21.2 | –19 35 | 9.9 | 3.0 |
| 6356 | | Oph | 17 23.6 | –17 49 | 8.4 | 7.2 |
| 6366 | | Oph | 17 27.7 | –05 05 | 10.0 | 8.3 |
| 6402 | M14 | Oph | 17 37.6 | –03 15 | 7.6 | 11.7 |
| 6517 | | Oph | 18 01.8 | –08 58 | 10.3 | 4.3 |
| 6535 | | Ser | 18 03.8 | –00 18 | 10.6 | 3.6 |
| 6539 | | Ser | 18 04.8 | –07 35 | 9.6 | 6.9 |
| I.1276 | | Ser | 18 10.7 | –07 12 | 10.3 | 7.1 |
| 6712 | | Sct | 18 53.1 | –08 42 | 8.2 | 7.2 |
| 6760 | | Aql | 19 11.2 | +01 02 | 9.1 | 6.6 |
| 6838 | M71 | Sge | 19 53.8 | +18 47 | 8.3 | 7.2 |

## Bright diffuse nebulae

| NGC/IC | Other | Con | RA | Dec | Type | Diam | Mag* | |
|---|---|---|---|---|---|---|---|---|
| | | | h m | ° ′ | | ′ | | |
| 6611 | M16 | Ser | 18 18.8 | –13 47 | E | 35 × 28 | — | Eagle Nebula, contains cluster |
| 6618 | M17 | Sgr | 18 20.8 | –16 11 | E | 46 × 37 | — | Omega Nebula, contains cluster |

## Planetary nebulae

| NGC/IC | Other | Con | RA | Dec | Mag | Diam | Mag* |
|---|---|---|---|---|---|---|---|
| | | | h m | ° ′ | p | ″ | |
| I.4593 | | Her | 16 12.2 | +12 04 | 10.9 | 12/120 | 11.3 |
| 6309 | | Oph | 17 14.1 | –12 55 | 10.8 | 14/66 | 14.4 |
| 6537 | | Sgr | 18 05.2 | –19 51 | 12.5 | 9 | — |
| 6572 | | Oph | 18 12.1 | +06 51 | 9.0 | 8 | 13.6 |
| 6741 | | Aql | 19 02.6 | –00 27 | 10.8 | 6 | 14.7 |
| 6751 | | Aql | 19 05.9 | –06 00 | 12.5 | 20 | 13.9 |
| 6781 | | Aql | 19 18.4 | +06 33 | 11.8 | 109 | 15.0 |
| 6790 | | Aql | 19 23.2 | +01 31 | 10.2 | 7 | 13.5 |
| 6803 | | Aql | 19 31.3 | +10 03 | 11.3 | 6 | 15.2 |
| — | PK52–2.2 | Aql | 19 39.2 | +15 57 | 12.6 | 10 | — |
| 6818 | | Sgr | 19 44.0 | –14 09 | 9.9 | 17 | 13.1 |

## Galaxies

| NGC/IC | Other | Con | RA | Dec | Mag | Size | Type |
|---|---|---|---|---|---|---|---|
| | | | h m | ° ′ | | ′ | |
| 6118 | | Ser | 16 21.8 | –02 17 | 12.3 | 4.7 × 2.3 | Sb |
| 6384 | | Oph | 17 32.4 | +07 04 | 10.6 | 6.0 × 4.3 | Sb |
| 6822 | | Sgr | 19 44.9 | –14 48 | 9.4 | 10.2 × 9.5 | Irr |

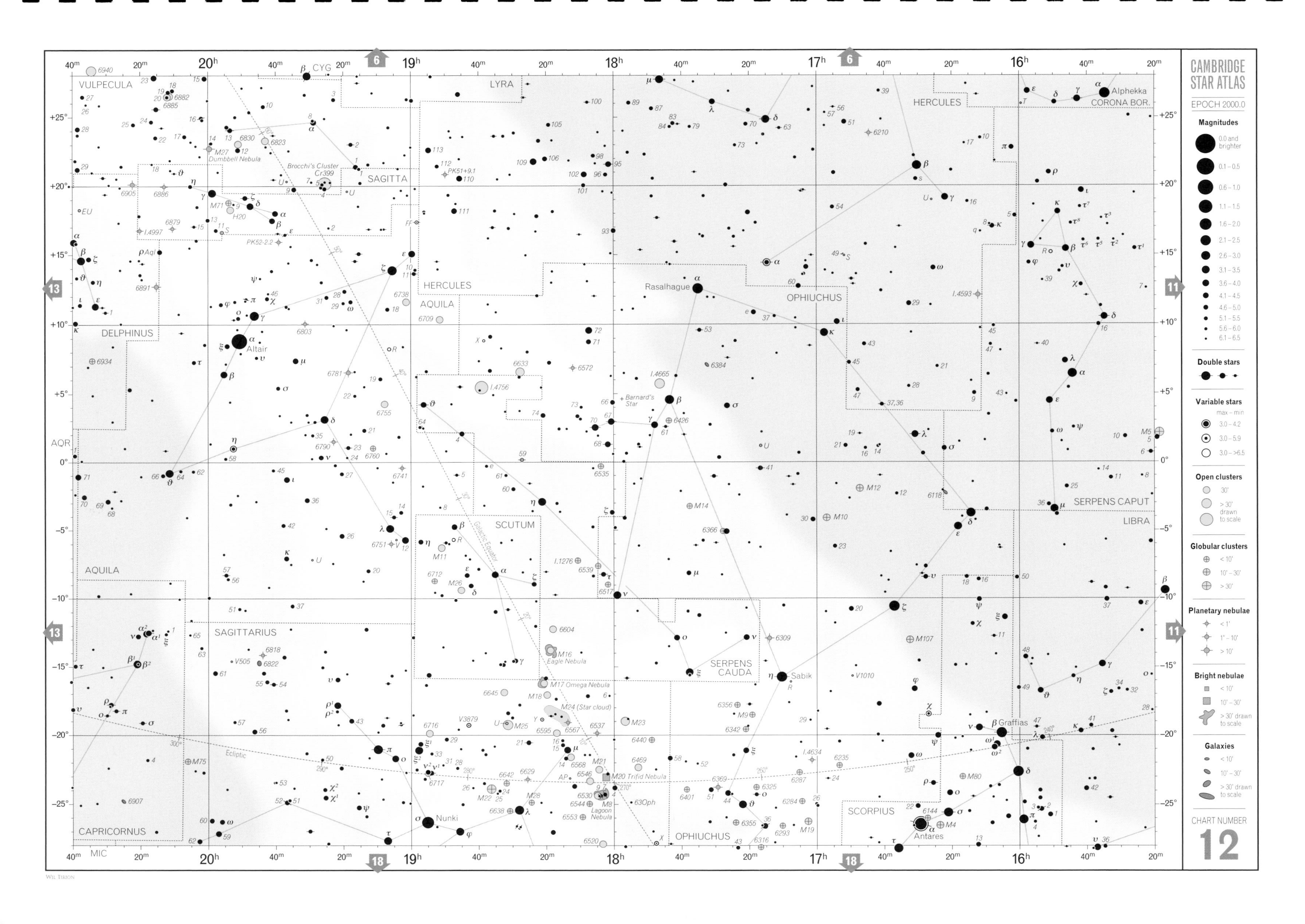
CAMBRIDGE STAR ATLAS
EPOCH 2000.0
Magnitudes
0.0 and brighter
0.1 – 0.5
0.6 – 1.0
1.1 – 1.5
1.6 – 2.0
2.1 – 2.5
2.6 – 3.0
3.1 – 3.5
3.6 – 4.0
4.1 – 4.5
4.6 – 5.0
5.1 – 5.5
5.6 – 6.0
6.1 – 6.5
Double stars
Variable stars
max – min
3.0 – 4.2
3.0 – 5.9
3.0 – >6.5
Open clusters
30'
> 30'
drawn to scale
Globular clusters
< 10'
10' – 30'
> 30'
Planetary nebulae
< 1'
1' – 10'
> 10'
Bright nebulae
< 10'
10' – 30'
> 30' drawn to scale
Galaxies
< 10'
10' – 30'
> 30' drawn to scale
CHART NUMBER
12
VULPECULA
CYG
SAGITTA
LYRA
HERCULES
CORONA BOR.
DELPHINUS
AQUILA
OPHIUCHUS
SERPENS CAPUT
LIBRA
SCUTUM
SERPENS CAUDA
SAGITTARIUS
CAPRICORNUS
SCORPIUS
AQR
MIC
Altair
Rasalhague
Alphekka
Sabik
Graffias
Antares
Nunki
Barnard's Star
Brocchi's Cluster
Cr399
Dumbbell Nebula
Eagle Nebula
M17 Omega Nebula
M24 (Star cloud)
M20 Trifid Nebula
Lagoon Nebula
Galactic Equator
Ecliptic
20h
19h
18h
17h
16h
+25°
+20°
+15°
+10°
+5°
0°
-5°
-10°
-15°
-20°
-25°
6
11
13
18
WIL TIRION

## Variable stars

| Star | Con | RA h m | Dec ° ′ | Range | Type | Period (days) | Spectrum |
|---|---|---|---|---|---|---|---|
| EU | Del | 20 37.9 | +18 16 | 5.8–6.9 | SR | 59.5 | M |
| U | Del | 20 45.5 | +15 05 | 5.6–7.5 | SR | 110: | M |
| EP | Aqr | 21 46.5 | –02 13 | 6.4–6.8 | SR | 55: | M |
| AG | Peg | 21 51.0 | +12 38 | 6.0–9.4 | ZA | 830.1 | M |
| R | Aqr | 23 43.8 | –15 17 | 5.8–12.4 | M | 387.0 | M |
| TX | Psc | 23 46.4 | +03 29 | 4.8–5.2 | Irr | — | M |

## Double stars

| Star | Con | RA h m | Dec ° ′ | PA ° | Sep ″ | Magnitudes | |
|---|---|---|---|---|---|---|---|
| 15 | Sge | 20 04.1 | +17 04 | 276 | 190.7 | 5.9 + 9.1 | |
| | | | | 320 | 203.7 | 6.8 | |
| α¹ | Cap | 20 17.6 | –12 30 | 221 | 45.4 | 4.2 + 9.2 | |
| α² | Cap | 20 18.1 | –12 33 | 172 | 6.6 | 3.6 + 10.4 | Algedi |
| α², α¹ | Cap | 20 18.1 | –12 33 | 291 | 377.7 | 3.6 + 4.2 | |
| σ | Cap | 20 19.4 | –19 07 | 179 | 55.9 | 5.5 + 9.0 | |
| π | Cap | 20 27.3 | –18 13 | 148 | 3.2 | 5.3 + 8.9 | |
| ρ | Cap | 20 28.9 | –17 49 | 158 | 0.5 | 5.0 + 10.0 | |
| 1 | Del | 20 30.3 | +10 54 | 346 | 0.9 | 6.1 + 8.1 | |
| γ | Del | 20 46.7 | +16 07 | 268 | 9.6 | 4.5 + 5.5 | |
| 13 | Del | 20 47.8 | +06 00 | 194 | 1.6 | 5.6 + 9.2 | |
| 1 | Equ | 20 59.1 | +04 18 | 284 | 0.8 | 6.0 + 6.3 | = ε Eql; binary 101.4 years |
| γ | Equ | 21 10.3 | +10 08 | 268 | 1.9 | 4.7 + 11.5 | |
| | | | | 5 | 47.7 | 12.5 | |
| | | | | 153 | 352.5 | 5.9 | |
| ζ | Aqr | 22 28.8 | –00 01 | 192 | 2.1 | 4.3 + 4.5 | Binary, 856 years |
| | | | | 191 | 41.0 | 7.5 | |
| 37 | Peg | 22 30.0 | +04 26 | 118 | 0.7 | 5.8 + 7.1 | Binary, 140 years |
| 107 | Aqr | 23 46.0 | –18 41 | 136 | 6.6 | 5.7 + 6.7 | |

## Open cluster

| NGC/IC | Other | Con | RA h m | Dec ° ′ | Mag | Diam ′ | N* | |
|---|---|---|---|---|---|---|---|---|
| 6994 | M73 | Aqr | 20 58.9 | –12 38 | 8.9 | 2.8 | 4 | Not a real cluster |

## Globular clusters

| NGC/IC | Other | Con | RA h m | Dec ° ′ | Mag | Diam ′ |
|---|---|---|---|---|---|---|
| 6934 | | Del | 20 34.2 | +07 24 | 8.9 | 5.9 |
| 6981 | M72 | Aqr | 20 53.5 | –12 32 | 9.4 | 5.9 |
| 7006 | | Del | 21 01.5 | +16 11 | 10.6 | 2.8 |
| 7078 | M15 | Peg | 21 30.0 | +12 10 | 6.4 | 12.3 |
| 7089 | M2 | Aqr | 21 33.5 | –00 49 | 6.5 | 12.9 |
| 7492 | | Aqr | 23 08.4 | –15 37 | 11.5 | 6.2 |

## Planetary nebulae

| NGC/IC | Other | Con | RA h m | Dec ° ′ | Mag p | Diam ″ | Mag* | |
|---|---|---|---|---|---|---|---|---|
| 6879 | | Sge | 20 10.5 | +16 55 | 13.0 | 5 | 15: | |
| 6886 | | Sge | 20 12.7 | +19 59 | 12.2 | 4 | 15.7 | |
| 6891 | | Del | 20 15.2 | +12 42 | 11.7 | 12/74 | 12.4 | |
| I.4997 | | Sge | 20 20.2 | +16 45 | 11.6 | 2 | 13: | |
| 7009 | | Aqr | 21 04.2 | –11 22 | 8.3 | 25/100 | 11.5 | Saturn Nebula |

## Galaxies

| NGC/IC | Other | Con | RA h m | Dec ° ′ | Mag | Size ′ | Type |
|---|---|---|---|---|---|---|---|
| 7448 | | Peg | 23 00.1 | +15 59 | 11.7 | 2.7 × 1.3 | Sc |
| 7479 | | Peg | 23 04.9 | +12 19 | 11.0 | 4.1 × 3.2 | SBb |
| 7606 | | Aqr | 23 19.1 | –08 29 | 10.8 | 5.8 × 2.6 | Sb |
| 7723 | | Aqr | 23 38.9 | –12 58 | 11.1 | 3.6 × 2.6 | Sb |
| 7727 | | Aqr | 23 39.9 | –12 18 | 10.7 | 4.2 × 3.4 | SBa |

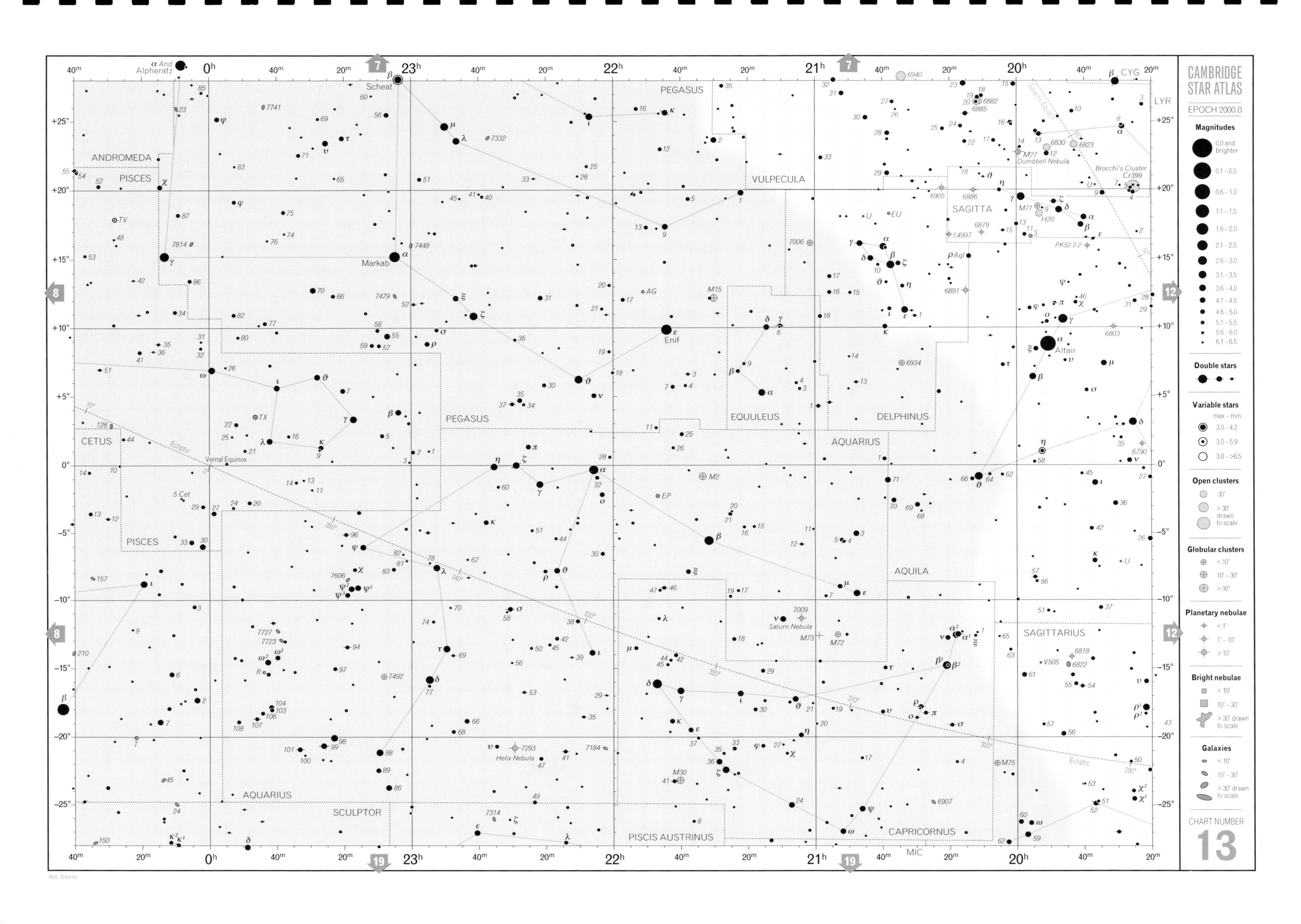

CAMBRIDGE STAR ATLAS
EPOCH 2000.0
Magnitudes
0.0 and brighter
0.1 – 0.5
0.6 – 1.0
1.1 – 1.5
1.6 – 2.0
2.1 – 2.5
2.6 – 3.0
3.1 – 3.5
3.6 – 4.0
4.1 – 4.5
4.6 – 5.0
5.1 – 5.5
5.6 – 6.0
6.1 – 6.5
Double stars
Variable stars
max – min
3.0 – 4.2
3.0 – 5.9
3.0 – >6.5
Open clusters
30'
> 30' drawn to scale
Globular clusters
< 10'
10' – 30'
> 30'
Planetary nebulae
< 1'
1' – 10'
> 10'
Bright nebulae
< 10'
10' – 30'
> 30' drawn to scale
Galaxies
< 10'
10' – 30'
> 30' drawn to scale
CHART NUMBER
13
ANDROMEDA
PISCES
PEGASUS
VULPECULA
SAGITTA
LYR
CYG
EQUULEUS
DELPHINUS
AQUARIUS
AQUILA
SAGITTARIUS
CETUS
SCULPTOR
PISCIS AUSTRINUS
CAPRICORNUS
MIC
Alpheratz
Scheat
Markab
Enif
Altair
Vernal Equinox
Ecliptic
Galactic Equator
Saturn Nebula
Helix Nebula
Dumbbell Nebula
Brocchi's Cluster
Cr399
M27
M71
M15
M2
M72
M73
M30
M75
H20
Wil Tirion

## Chart 14 *RA 0ʰ to 4ʰ, declination −20° to −65°*

### Variable stars

| Star | Con | RA h m | Dec ° ′ | Range | Type | Period (days) | Spectrum |
|---|---|---|---|---|---|---|---|
| S | Scl | 00 15.4 | −32 03 | 5.5–13.6 | M | 365.3 | M |
| T | Cet | 00 21.8 | −20 03 | 5.0–6.9 | SR | 158.9 | M |
| ζ | Phe | 01 08.4 | −55 15 | 3.9–4.4 | EA | 1.67 | B |
| R | Scl | 01 27.0 | −32 33 | 9.1–12.8 p | SR | 370 | M |
| AA | Cet | 01 59.0 | −22 55 | 6.0–6.5 | EW | 0.54 | F |
| R | Hor | 02 53.9 | −49 53 | 4.7–14.3 | M | 404.0 | M |
| V | Hor | 03 03.5 | −58 56 | 8.7–9.8 p | SR | — | M |
| TW | Hor | 03 12.6 | −57 19 | 5.3–6.0 | SR | 158: | K |

### Double stars

| Star | Con | RA h m | Dec ° ′ | PA ° | Sep ″ | Magnitudes | |
|---|---|---|---|---|---|---|---|
| $\kappa^1$ | Scl | 00 09.3 | −27 59 | 265 | 1.4 | 6.1 + 6.2 | |
| β | Tuc | 00 31.5 | −62 58 | 169 | 27.1 | 4.4 + 4.8 | |
| ξ | Phe | 00 41.8 | −56 30 | 253 | 13.2 | 5.8 + 10.2 | |
| $\lambda^1$ | Scl | 00 42.7 | −38 28 | 3 | 0.5 | 6.7 + 7.0 | |
| β | Phe | 01 06.1 | −46 43 | 346 | 1.4 | 4.0 + 4.2 | |
| p | Eri | 01 39.8 | −56 12 | 191 | 11.5 | 5.8 + 5.8 | Binary, 483.7 years |
| ε | Scl | 01 45.6 | −25 03 | 23 | 4.7 | 5.4 + 8.6 | Binary, 1192 years |
| ω | For | 02 33.8 | −28 14 | 244 | 10.8 | 5.0 + 7.7 | |
| $\eta^2$ | For | 02 50.2 | −35 51 | 14 | 5.0 | 5.9 + 10.1 | |
| ϑ | Eri | 02 58.3 | −40 18 | 88 | 8.2 | 3.4 + 4.5 | |
| α | For | 03 12.1 | −28 59 | 299 | 5.1 | 4.0 + 7.0 v | Binary, 314 years |
| $\tau^4$ | Eri | 03 19.5 | −21 45 | 288 | 5.7 | 3.7 + 9.2 | |
| | | | | 236 | 160.2 | 9.7 | |
| $\chi^3$ | For | 03 28.2 | −35 51 | 248 | 6.3 | 6.5 + 10.5 | |

### Globular clusters

| NGC/IC | Other | Con | RA h m | Dec ° ′ | Mag | Diam ′ |
|---|---|---|---|---|---|---|
| 288 | | Scl | 00 52.8 | −26 35 | 8.1 | 13.8 |
| 1261 | | Hor | 03 12.3 | −55 13 | 8.4 | 6.9 |

### Planetary nebula

| NGC/IC | Other | Con | RA h m | Dec ° ′ | Mag p | Diam ″ | Mag* |
|---|---|---|---|---|---|---|---|
| 1360 | | For | 03 33.3 | −25 51 | 9.4 | 390 | 11.4 |

### Galaxies

| NGC/IC | Other | Con | RA h m | Dec ° ′ | Mag | Size ′ | Type | |
|---|---|---|---|---|---|---|---|---|
| 24 | | Scl | 00 09.9 | −24 58 | 11.5 | 5.5 × 1.6 | Sb | |
| 45 | | Cet | 00 14.1 | −23 11 | 10.4 | 8.1 × 5.8 | S | |
| 55 | | Scl | 00 14.9 | −39 11 | 8.2 | 32.4 × 6.5 | SB | |
| 134 | | Scl | 00 30.4 | −33 15 | 10.1 | 8.1 × 2.6 | SBb | |
| 150 | | Scl | 00 34.3 | −27 48 | 11.1 | 4.2 × 2.3 | S | |
| 247 | | Cet | 00 47.1 | −20 46 | 8.9 | 20.0 × 7.4 | Sc | |
| 253 | | Scl | 00 47.6 | −25 17 | 7.1 | 25.1 × 7.4 | Sc | |
| 289 | | Scl | 00 52.7 | −31 12 | 11.6 | 3.7 × 2.7 | Sb | |
| 300 | | Scl | 00 54.9 | −37 41 | 8.7 | 20.0 × 14.8 | Sd | |
| — | E 351–30 | Scl | 00 59.9 | −33 42 | 10.5 | 35: × 29: | dE3 | Sculptor Dwarf |
| 578 | | Cet | 01 30.5 | −22 40 | 10.9 | 4.8 × 3.2 | Sc | |
| 613 | | Scl | 01 34.3 | −29 25 | 10.0 | 5.8 × 4.6 | SBb | |
| 685 | | Eri | 01 47.8 | −52 47 | 11.8 | 4.1 × 4.0 | SBc | |
| 908 | | Cet | 02 23.1 | −21 14 | 10.2 | 5.5 × 2.8 | SBb | |
| 986 | | For | 02 33.6 | −39 02 | 11.0 | 3.7 × 2.8 | Sc | |
| — | E 356–4 | For | 02 39.9 | −34 32 | 9.0 | 63: × 48: | dE3 | Fornax Dwarf |
| 1097 | | For | 02 46.3 | −30 17 | 9.3 | 9.3 × 6.6 | SBb | |
| 1187 | | Eri | 03 02.6 | −22 52 | 10.9 | 5.0 × 4.0 | SBc | |
| 1201 | | For | 03 04.1 | −26 04 | 10.6 | 4.4 × 2.8 | Sa | |
| 1232 | | Eri | 03 09.8 | −20 35 | 9.9 | 7.8 × 6.9 | Sc | |
| 1249 | | Hor | 03 10.1 | −53 21 | 11.7 | 5.2 × 2.7 | SBc | |
| 1255 | | For | 03 13.5 | −25 44 | 11.1 | 4.1 × 2.8 | Sc | |
| 1291 | | Eri | 03 17.3 | −41 08 | 8.5 | 10.5 × 9.1 | SBa | |
| 1302 | | For | 03 19.9 | −26 04 | 11.5 | 4.4 × 4.2 | SBa | |
| 1316 | | For | 03 22.7 | −37 12 | 8.9 | 7.1 × 5.5 | SB0 | |
| 1326 | | For | 03 23.9 | −36 28 | 10.5 | 4.0 × 3.0 | SB0 | |
| 1332 | | Eri | 03 26.3 | −21 20 | 10.3 | 4.6 × 1.7 | E7 | |
| 1344 | | For | 03 28.3 | −31 04 | 10.3 | 3.9 × 2.3 | E3 | |
| 1350 | | For | 03 31.1 | −33 38 | 10.5 | 4.3 × 2.4 | SBb | |
| 1365 | | For | 03 33.6 | −36 08 | 9.5 | 9.8 × 5.5 | SBb | |
| 1371 | | For | 03 35.0 | −24 56 | 11.5 | 5.4 × 4.0 | SBa | |
| 1380 | | For | 03 36.5 | −34 59 | 11.1 | 4.9 × 1.9 | S0 | |
| 1385 | | For | 03 37.5 | −24 30 | 11.2 | 3.0 × 2.0 | Sc | |
| 1395 | | Eri | 03 38.5 | −23 02 | 11.3 | 3.2 × 2.5 | E3 | |
| 1398 | | For | 03 38.9 | −26 20 | 9.7 | 6.6 × 5.2 | SBb | |
| 1399 | | For | 03 38.5 | −35 27 | 9.9 | 3.2 × 3.1 | E1p | |
| 1411 | | Hor | 03 38.8 | −44 05 | 11.9 | 2.8 × 2.3 | S0 | |
| 1404 | | For | 03 38.9 | −35 35 | 10.3 | 2.5 × 2.3 | E1 | |
| 1433 | | Hor | 03 42.0 | −47 13 | 10.0 | 6.8 × 6.0 | SBa | |
| 1425 | | For | 03 42.2 | −29 54 | 11.7 | 5.4 × 2.7 | Sb | |
| 1448 | | Hor | 03 44.5 | −44 39 | 11.3 | 8.1 × 1.8 | Sc | |
| 1493 | | Hor | 03 57.5 | −46 12 | 11.8 | 2.6 × 2.3 | SBc | |

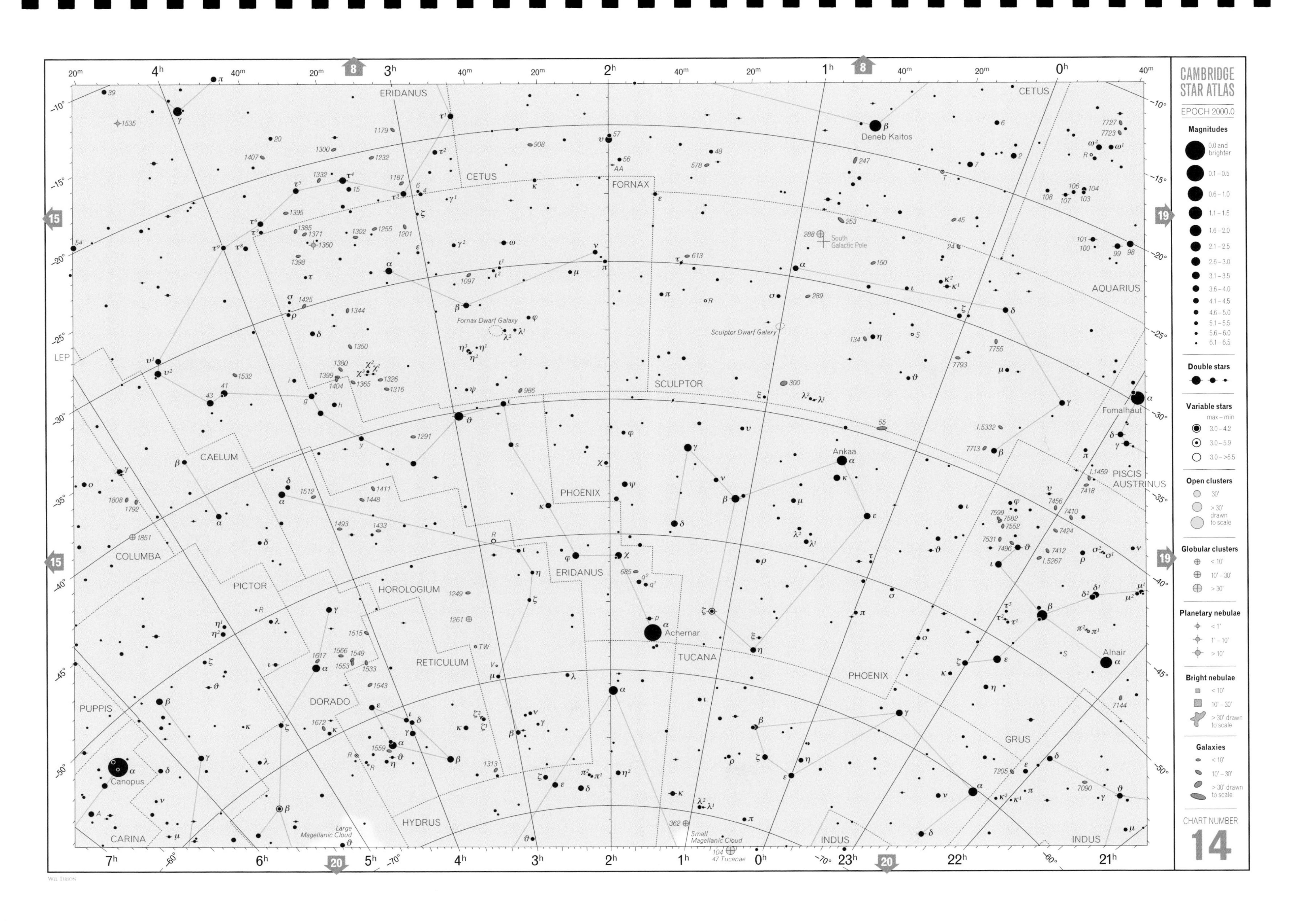
CAMBRIDGE STAR ATLAS
EPOCH 2000.0
Magnitudes
0.0 and brighter
0.1 – 0.5
0.6 – 1.0
1.1 – 1.5
1.6 – 2.0
2.1 – 2.5
2.6 – 3.0
3.1 – 3.5
3.6 – 4.0
4.1 – 4.5
4.6 – 5.0
5.1 – 5.5
5.6 – 6.0
6.1 – 6.5
Double stars
Variable stars
max – min
3.0 – 4.2
3.0 – 5.9
3.0 – >6.5
Open clusters
30'
> 30' drawn to scale
Globular clusters
< 10'
10' – 30'
> 30'
Planetary nebulae
< 1'
1' – 10'
> 10'
Bright nebulae
< 10'
10' – 30'
> 30' drawn to scale
Galaxies
< 10'
10' – 30'
> 30' drawn to scale
CHART NUMBER
14
ERIDANUS
CETUS
FORNAX
SCULPTOR
AQUARIUS
PISCIS AUSTRINUS
PHOENIX
CAELUM
COLUMBA
PICTOR
HOROLOGIUM
RETICULUM
DORADO
TUCANA
GRUS
HYDRUS
INDUS
PUPPIS
CARINA
LEP
Deneb Kaitos
Fomalhaut
Ankaa
Achernar
Alnair
Canopus
South Galactic Pole
Fornax Dwarf Galaxy
Sculptor Dwarf Galaxy
Large Magellanic Cloud
Small Magellanic Cloud
47 Tucanae
WIL TIRION

## Variable stars

| Star | Con | RA h m | Dec ° ′ | Range | Type | Period (days) | Spectrum |
|---|---|---|---|---|---|---|---|
| R | Ret | 04 33.5 | –63 02 | 6.5–14.0 | M | 278.3 | M |
| R | Dor | 04 36.8 | –62 05 | 4.8–6.6 | SR | 338: | M |
| R | Pic | 04 46.2 | –49 15 | 6.7–10.0 | SR | 164.2 | M |
| β | Dor | 05 33.6 | –62 29 | 3.5–4.1 | Cep | 9.84 | F–G |
| S | Lep | 06 05.8 | –24 12 | 6.0–7.6 | SR | 90 | B |
| L² | Pup | 07 13.5 | –44 39 | 2.6–6.2 | SR | 140.4 | M |
| ω | CMa | 07 14.8 | –26 46 | 3.6–4.2 | Irr | — | B |
| V | Pup | 07 58.2 | –49 15 | 4.4–4.9 | EB | 1.45 | B + B |

## Double stars

| Star | Con | RA h m | Dec ° ′ | PA ° | Sep ″ | Magnitudes | |
|---|---|---|---|---|---|---|---|
| γ | Cae | 05 40.4 | –25 29 | 308 | 2.9 | 4.6 + 8.1 | |
| ϑ | Pic | 05 24.8 | –52 19 | 287 | 38.2 | 6.3 + 6.8 | |
| β | Lep | 05 28.2 | –20 46 | 330 | 2.5 | 2.8 + 7.3 | Nihal |
| γ | Lep | 05 44.5 | –22 27 | 350 | 96.3 | 3.7 + 6.3 | |
| μ | Pic | 06 32.0 | –58 45 | 231 | 2.4 | 5.8 + 9.0 | |
| 17 | CMa | 06 55.0 | –20 24 | 147 | 44.4 | 5.8 + 9.3 | |
| π | CMa | 06 55.6 | –20 08 | 18 | 11.6 | 4.7 + 9.7 | |
| ε | CMa | 06 58.6 | –28 58 | 161 | 7.5 | 1.5 + 7.4 | Adhara |
| σ | Pup | 07 29.2 | –43 18 | 74 | 22.3 | 3.3 + 9.4 | |

## Open clusters

| NGC/IC | Other | Con | RA h m | Dec ° ′ | Mag | Diam ′ | N* | |
|---|---|---|---|---|---|---|---|---|
| 2287 | M41 | CMa | 06 47.0 | –20 44 | 4.5 | 38 | 80 | |
| 2354 | | CMa | 07 14.3 | –25 44 | 6.5 | 20 | 100 | |
| 2362 | | CMa | 07 18.8 | –24 57 | 4.1 | 8 | 60 | |
| — | Cr 135 | Pup | 07 17.0 | –36 50 | 2.1 | 50 | — | Contains π Pup |
| 2367 | | CMa | 07 20.1 | –21 56 | 7.9 | 3.5 | 30 | |
| 2383 | | CMa | 07 24.8 | –20 56 | 8.4 | 6 | 40 | |
| 2384 | | CMa | 07 25.1 | –21 02 | 7.4 | 2.5 | 15 | |
| 2421 | | Pup | 07 36.3 | –20 37 | 8.3 | 10 | 70 | |
| 2439 | | Pup | 07 40.8 | –31 39 | 6.9 | 10 | 80 | |
| 2447 | M93 | Pup | 07 44.6 | –23 52 | 6.2 | 22 | 80 | |
| 2451 | | Pup | 07 45.4 | –37 58 | 2.8 | 45 | 40 | Contains c Pup |
| 2453 | | Pup | 07 47.8 | –27 14 | 8.3 | 5 | 30 | |
| 2477 | | Pup | 07 52.3 | –38 33 | 5.8 | 27 | 160 | |
| 2467 | | Pup | 07 52.6 | –26 23 | 7.1 | 16 | 50 | Contains nebula |
| 2482 | | Pup | 07 54.9 | –24 18 | 7.3 | 12 | 40 | |
| 2489 | | Pup | 07 56.2 | –30 04 | 7.9 | 8 | 45 | |
| 2516 | | Car | 07 58.3 | –60 53 | 3.8 | 30 | 80 | |

## Globular clusters

| NGC/IC | Other | Con | RA h m | Dec ° ′ | Mag | Diam ′ |
|---|---|---|---|---|---|---|
| 1851 | | Col | 05 14.1 | –40 03 | 7.3 | 11.0 |
| 1904 | M79 | Lep | 05 24.5 | –24 33 | 8.0 | 8.7 |
| 2298 | | Pup | 06 49.0 | –36 00 | 9.4 | 6.8 |

## Bright diffuse nebula

| NGC/IC | Other | Con | RA h m | Dec ° ′ | Type | Diam ′ | Mag* | |
|---|---|---|---|---|---|---|---|---|
| 2467 | | Pup | 07 52.6 | –26 24 | E | 8 × 7 | 9.2 | In cluster |

## Galaxies

| NGC/IC | Other | Con | RA h m | Dec ° ′ | Mag | Size ′ | Type |
|---|---|---|---|---|---|---|---|
| 1512 | | Hor | 04 03.9 | –43 21 | 10.6 | 4.0 × 3.2 | SBa |
| 1515 | | Dor | 04 04.1 | –54 06 | 11.0 | 5.4 × 1.3 | SBb |
| 1532 | | Eri | 04 12.1 | –32 52 | 11.1 | 5.6 × 1.8 | Sb |
| 1543 | | Ret | 04 12.8 | –57 44 | 10.6 | 3.9 × 2.1 | SB0 |
| 1533 | | Dor | 04 16.2 | –55 47 | 9.5 | 4.1 × 2.8 | S0 |
| 1549 | | Dor | 04 15.7 | –55 36 | 9.9 | 3.7 × 3.2 | E0 |
| 1553 | | Dor | 04 16.2 | –55 47 | 9.5 | 4.1 × 2.3 | S0 |
| 1559 | | Ret | 04 17.6 | –62 47 | 10.5 | 3.3 × 2.1 | SBc |
| 1566 | | Dor | 04 20.0 | –54 56 | 9.4 | 7.6 × 6.2 | SBb |
| 1617 | | Dor | 04 31.7 | –54 36 | 10.4 | 4.7 × 2.4 | SBa |
| 1672 | | Dor | 04 45.7 | –59 15 | 11.0 | 4.8 × 3.9 | SBb |
| 1744 | | Lep | 05 00.0 | –26 01 | 11.2 | 6.8 × 4.1 | SBc |
| 1792 | | Col | 05 05.2 | –37 59 | 10.2 | 4.0 × 2.1 | Sb |
| 1808 | | Col | 05 07.7 | –37 31 | 9.9 | 7.2 × 4.1 | SBa |
| 1964 | | Lep | 05 33.4 | –21 57 | 10.8 | 6.2 × 2.5 | Sb |
| 2090 | | Col | 05 47.0 | –34 14 | 11.8 | 4.5 × 2.3 | Sc |
| 2207 | | CMa | 06 16.4 | –21 22 | 10.7 | 4.3 × 2.9 | Sc |
| 2217 | | CMa | 06 21.7 | –27 14 | 10.4 | 4.8 × 4.4 | SBa |
| 2223 | | CMa | 06 24.6 | –22 50 | 11.4 | 3.3 × 3.0 | SBb |
| 2280 | | CMa | 06 44.8 | –27 38 | 11.8 | 5.6 × 3.2 | Sb |

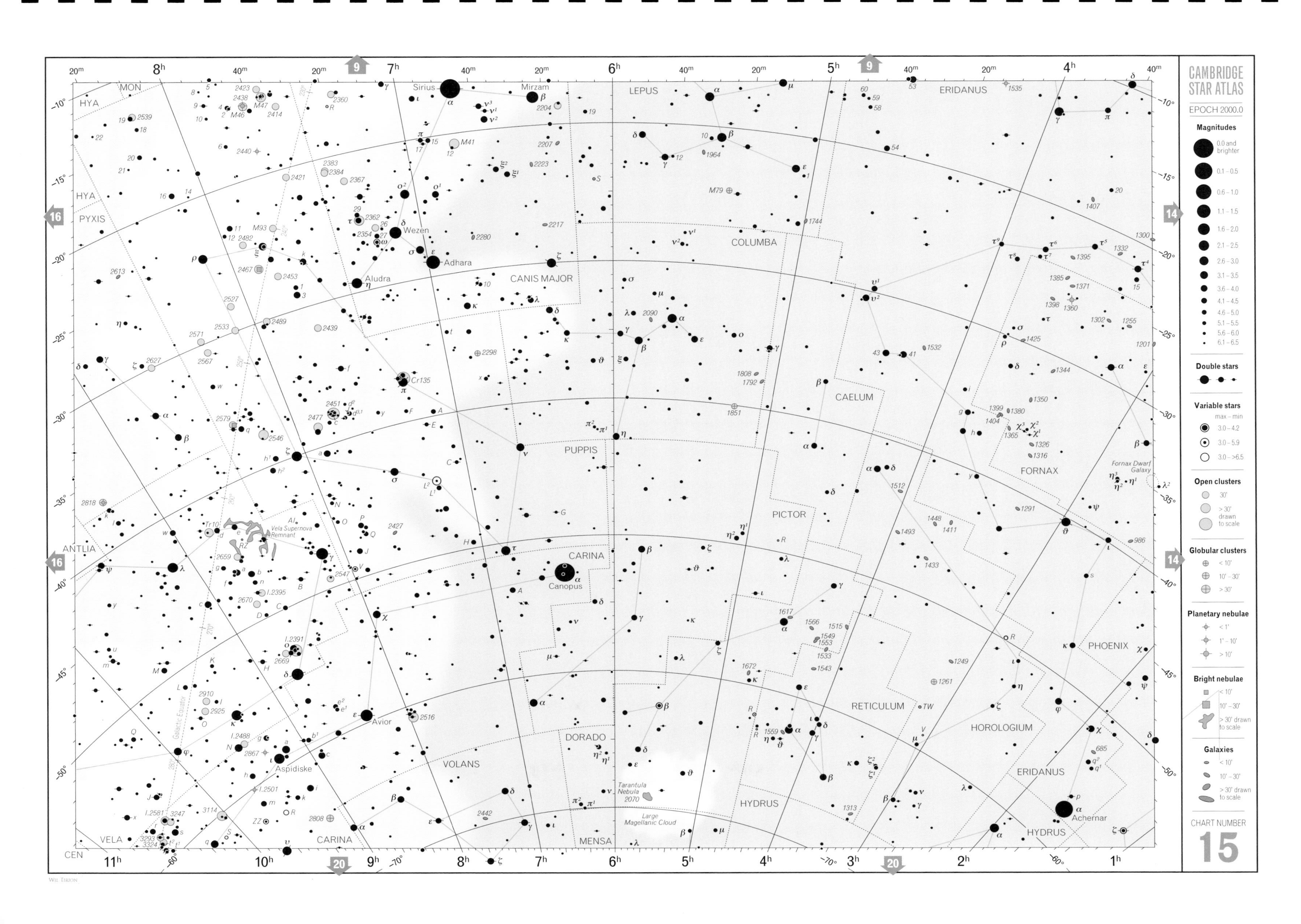

CAMBRIDGE STAR ATLAS
EPOCH 2000.0
Magnitudes
0.0 and brighter
0.1 – 0.5
0.6 – 1.0
1.1 – 1.5
1.6 – 2.0
2.1 – 2.5
2.6 – 3.0
3.1 – 3.5
3.6 – 4.0
4.1 – 4.5
4.6 – 5.0
5.1 – 5.5
5.6 – 6.0
6.1 – 6.5
Double stars
Variable stars
max – min
3.0 – 4.2
3.0 – 5.9
3.0 – >6.5
Open clusters
30'
> 30' drawn to scale
Globular clusters
< 10'
10' – 30'
> 30'
Planetary nebulae
< 1'
1' – 10'
> 10'
Bright nebulae
< 10'
10' – 30'
> 30' drawn to scale
Galaxies
< 10'
10' – 30'
> 30' drawn to scale
CHART NUMBER
15
LEPUS
ERIDANUS
COLUMBA
CANIS MAJOR
CAELUM
FORNAX
PUPPIS
PICTOR
CARINA
PHOENIX
HOROLOGIUM
RETICULUM
DORADO
VOLANS
HYDRUS
MENSA
VELA
ANTLIA
PYXIS
HYA
MON
CEN
Sirius
Mirzam
Wezen
Adhara
Aludra
Canopus
Avior
Aspidiske
Achernar
Vela Supernova Remnant
Fornax Dwarf Galaxy
Tarantula Nebula
Large Magellanic Cloud
Galactic Equator
Wil Tirion

## Variable stars

| Star | Con | RA h m | Dec ° ′ | Range | Type | Period (days) | Spectrum |
|---|---|---|---|---|---|---|---|
| AI | Vel | 08 14.1 | −44 34 | 6.4–7.1 | δSct | 0.11 | A–F |
| RZ | Vel | 08 37.0 | −44 07 | 6.4–7.6 | Cep | 20.39 | G |
| R | Car | 09 32.2 | −62 47 | 3.9–10.5 | M | 308.7 | M |
| ZZ | Car | 09 45.2 | −62 30 | 3.3–4.2 | Cep | 35.53 | F–K |
| Y | Hya | 09 51.1 | −23 01 | 8.3–12.0 p | SR | 302.8 | K |
| S | Car | 10 09.4 | −61 33 | 4.5–9.9 | M | 149.5 | K–M |
| U | Ant | 10 35.2 | −39 34 | 8.1–9.7 p | Irr | — | K |
| η | Car | 10 45.1 | −59 41 | −0.8–7.9 | SD | — | Pec |
| U | Car | 10 57.8 | −59 44 | 5.7–7.0 | Cep | 38.76 | F–G |
| ER | Car | 11 09.7 | −58 50 | 6.6–7.1 | Cep | 7.72 | F–G |
| o¹ | Cen | 11 31.8 | −59 27 | 4.7–5.5 | SR | 200: | G |

## Double stars

| Star | Con | RA h m | Dec ° ′ | PA ° | Sep ″ | Magnitudes | |
|---|---|---|---|---|---|---|---|
| γ | Vel | 08 09.5 | −47 20 | 220 | 41.2 | 1.9 + 4.2 | |
| | | | | 151 | 62.3 | 8.3 | |
| | | | | 141 | 93.5 | 9.1 | |
| | | | | 146 | 1.8 | 12.5 | |
| δ | Vel | 08 44.7 | −54 43 | 153 | 2.6 | 2.1 + 5.1 | |
| H | Vel | 08 56.3 | −52 43 | 339 | 2.7 | 4.8 + 7.4 | |
| κ | Pyx | 09 08.0 | −25 52 | 263 | 2.1 | 4.6 + 9.8 | |
| ζ¹ | Ant | 09 30.8 | −31 53 | 212 | 8.0 | 6.2 + 7.1 | |
| δ | Ant | 10 29.6 | −30 36 | 226 | 11.0 | 5.6 + 9.6 | |
| μ | Vel | 10 46.8 | −49 25 | 58 | 2.5 | 2.7 + 6.4 | Binary, 116.2 years |
| β | Hya | 11 52.9 | −33 54 | 8 | 0.9 | 4.7 + 5.5 | |

## Open clusters

| NGC/IC | Other | Con | RA h m | Dec ° ′ | Mag | Diam ′ | N* | |
|---|---|---|---|---|---|---|---|---|
| 2527 | | Pup | 08 05.3 | −28 10 | 6.5 | 22 | 40 | |
| 2533 | | Pup | 08 07.0 | −29 54 | 7.6 | 3.5 | 60 | |
| 2547 | | Vel | 08 10.7 | −49 16 | 4.7 | 20 | 80 | |
| 2546 | | Pup | 08 12.4 | −37 38 | 6.3 | 41 | 40 | |
| 2567 | | Pup | 08 18.6 | −30 38 | 7.4 | 10 | 40 | |
| 2571 | | Pup | 08 18.9 | −29 44 | 7.0 | 13 | 30 | |
| 2579 | | Pup | 08 21.1 | −36 11 | 7.5 | 10 | 20 | |
| 2627 | | Pyx | 08 37.3 | −29 57 | 8.4 | 11 | 60 | |
| I.2391 | | Vel | 08 40.2 | −53 04 | 2.5 | 50 | 30 | o Velorum Cluster |
| I.2395 | | Vel | 08 41.1 | −48 12 | 4.6 | 8 | 40 | |
| 2669 | | Vel | 08 44.9 | −52 58 | 6.1 | 12 | 40 | |
| 2670 | | Vel | 08 45.5 | −48 47 | 7.8 | 9 | 30 | |
| — | Tr 10 | Vel | 08 47.8 | −42 29 | 4.6 | 15 | 40 | |
| 2818 | | Pyx | 09 16.0 | −36 37 | 8.2 | 9 | 40 | Cont. planetary neb. |
| I.2488 | | Vel | 09 27.6 | −56 59 | 7.4 | 15 | 70 | |
| 2910 | | Vel | 09 30.4 | −52 54 | 7.2 | 5 | 30 | |
| 2925 | | Vel | 09 33.7 | −53 26 | 8.3 | 12 | 40 | |
| 3114 | | Car | 10 02.7 | −60 07 | 4.2 | 35 | 100 | |
| 3247 | | Car | 10 25.9 | −57 56 | 7.6 | 7 | 20 | |
| I.2581 | | Car | 10 27.4 | −57 38 | 4.3 | 8 | 25 | |
| 3293 | | Car | 10 35.8 | −58 14 | 4.7 | 6 | 30 | |
| 3324 | | Car | 10 37.3 | −58 38 | 6.7 | 6 | 10 | |
| I.2602 | | Cep | 10 43.2 | −64 24 | 1.9 | 50 | 60 | Southern Pleiades |
| 3532 | | Car | 11 06.4 | −58 40 | 3.0 | 55 | 150 | |

## Open clusters *(continued)*

| NGC/IC | Other | Con | RA h m | Dec ° ′ | Mag | Diam ′ | N* | |
|---|---|---|---|---|---|---|---|---|
| 3572 | | Car | 11 10.4 | −60 14 | 6.6 | 7 | 35 | |
| 3590 | | Car | 11 12.9 | −60 47 | 8.2 | 4 | 25 | |
| I.2714 | | Car | 11 17.9 | −62 42 | 8.2 | 12 | 100 | |
| — | Mel 105 | Car | 11 19.5 | −63 30 | 70 | 4 | 70 | |
| 3680 | | Cen | 11 25.7 | −43 15 | 7.6 | 12 | 30 | |
| 3766 | | Cen | 11 36.1 | −61 37 | 5.3 | 12 | 100 | |
| I.2944 | | Cen | 11 36.6 | −63 02 | 4.5 | 15 | 30 | λ Centauri Cluster |
| 3960 | | Cen | 11 50.9 | −55 42 | 8.3 | 7 | 45 | |

## Globular clusters

| NGC/IC | Other | Con | RA h m | Dec ° ′ | Mag | Diam ′ |
|---|---|---|---|---|---|---|
| 2808 | | Car | 09 12.0 | −64 52 | 6.3 | 13.8 |
| 3201 | | Vel | 10 17.6 | −46 25 | 6.8 | 18.2 |

## Bright diffuse nebulae

| NGC/IC | Other | Con | RA h m | Dec ° ′ | Type | Diam ′ | Mag* | |
|---|---|---|---|---|---|---|---|---|
| 2579 | | Pup | 08 20.9 | −36 13 | E | 2 | 1.2 | |
| — | Gum 12 | Vel | 08 30: | −45: | E | 1200 × 720 | | Vela SNR |
| 3372 | | Car | 10 43.8 | −59 52 | E | 120 × 120 | 6.2 | Eta Carinae Nebula |

## Planetary nebulae

| NGC/IC | Other | Con | RA h m | Dec ° ′ | Mag p | Diam ″ | Mag* | |
|---|---|---|---|---|---|---|---|---|
| 2818 | | Pyx | 09 16.0 | −36 38 | 13.0 | 38 | | In open cluster |
| 2867 | | Car | 09 21.4 | −58 19 | 9.7 | 11 | 13.6 | |
| I.2501 | | Car | 09 38.8 | −60 05 | 11.3 | 25 | | |
| 3132 | | Vel | 10 07.7 | −40 26 | 8.2 | 47 | 10.1 | |
| 3211 | | Car | 10 17.8 | −62 40 | 11.8 | 12 | | |
| 3918 | | Cen | 11 50.3 | −57 11 | 8.4 | 12 | 10.8 | |

## Galaxies

| NGC/IC | Other | Con | RA h m | Dec ° ′ | Mag | Size ′ | Type |
|---|---|---|---|---|---|---|---|
| 2613 | | Pyx | 08 33.4 | −22 58 | 10.4 | 7.2 × 2.1 | Sb |
| 2784 | | Hya | 09 12.3 | −24 10 | 10.1 | 5.1 × 2.3 | S0 |
| 2835 | | Hya | 09 17.9 | −22 21 | 11.1 | 6.3 × 4.4 | S |
| 2935 | | Hya | 09 36.7 | −21 08 | 11.9 | 3.5 × 3.0 | SBb |
| 2997 | | Ant | 09 45.6 | −31 11 | 10.6 | 8.1 × 6.5 | Sc |
| 3109 | | Hya | 10 03.1 | −26 09 | 10.4 | 14.5 × 3.5 | Irr |
| 3223 | | Ant | 10 21.6 | −34 16 | 11.8 | 4.1 × 2.6 | Sb |
| 3256 | | Vel | 10 27.8 | −43 54 | 11.3 | 3.5 × 2.0 | Pec |
| 3309 | | Hya | 10 36.6 | −27 31 | 11.9 | 1.9 × 1.7 | E0 |
| 3511 | | Crt | 11 03.4 | −23 05 | 11.6 | 5.4 × 2.2 | Sc |
| 3513 | | Crt | 11 03.8 | −23 15 | 12.0 | 2.8 × 2.3 | SBc |
| 3557 | | Cen | 11 10.0 | −37 32 | 10.4 | 4.0 × 2.7 | E3 |
| 3585 | | Hya | 11 13.3 | −26 45 | 10.0 | 2.9 × 1.6 | E5 |
| 3621 | | Hya | 11 18.3 | −32 49 | 9.9 | 10.0 × 6.5 | Sc |
| 3904 | | Hya | 11 49.2 | −29 17 | 11.0 | 2.2 × 1.7 | E2 |
| 3923 | | Hya | 11 51.0 | −28 48 | 10.1 | 2.9 × 1.9 | E3 |

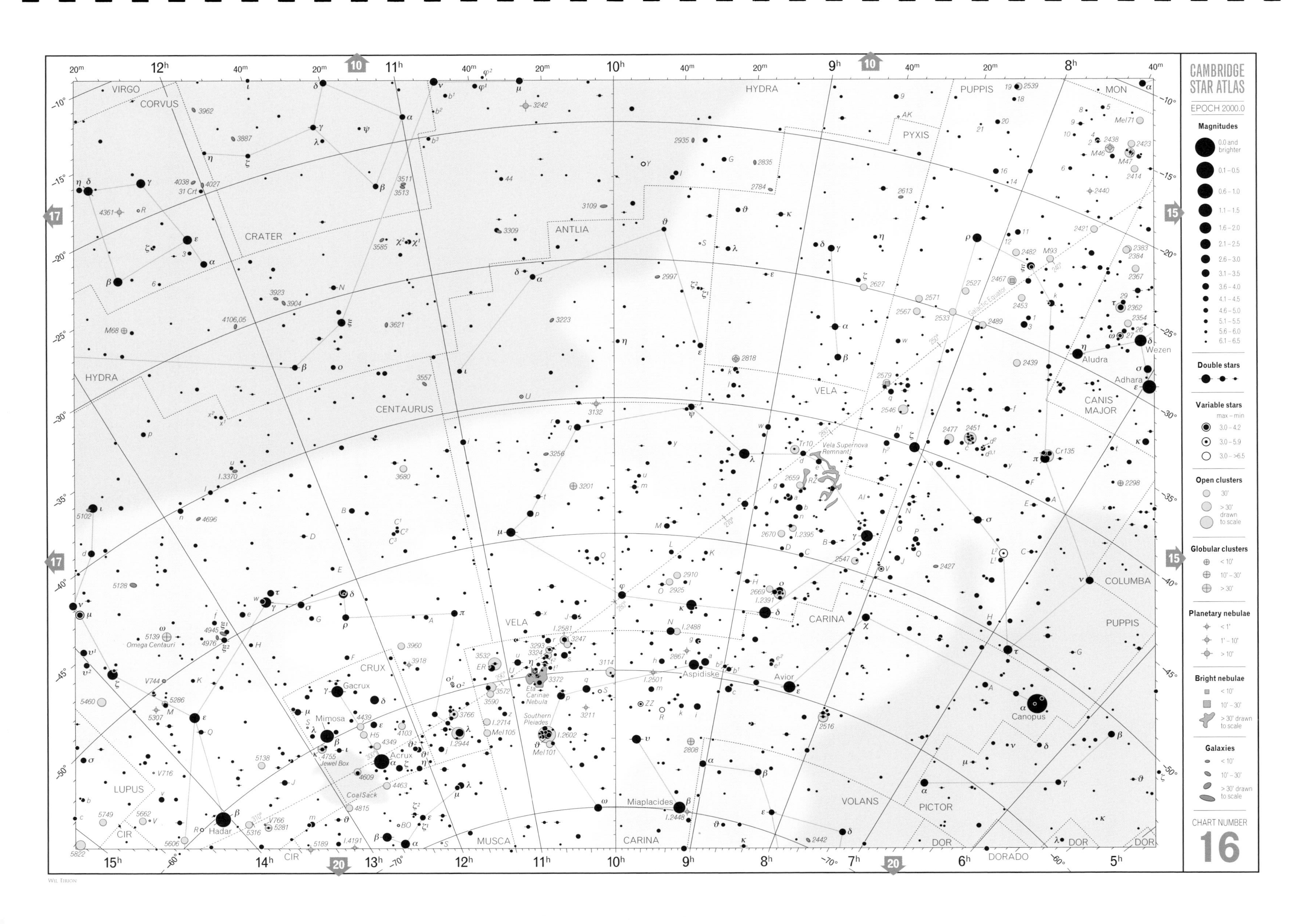
CAMBRIDGE STAR ATLAS
EPOCH 2000.0
Magnitudes
0.0 and brighter
0.1 – 0.5
0.6 – 1.0
1.1 – 1.5
1.6 – 2.0
2.1 – 2.5
2.6 – 3.0
3.1 – 3.5
3.6 – 4.0
4.1 – 4.5
4.6 – 5.0
5.1 – 5.5
5.6 – 6.0
6.1 – 6.5
Double stars
Variable stars
max – min
3.0 – 4.2
3.0 – 5.9
3.0 – >6.5
Open clusters
30'
> 30' drawn to scale
Globular clusters
< 10'
10' – 30'
> 30'
Planetary nebulae
< 1'
1' – 10'
> 10'
Bright nebulae
< 10'
10' – 30'
> 30' drawn to scale
Galaxies
< 10'
10' – 30'
> 30' drawn to scale
CHART NUMBER
16
VIRGO
CORVUS
CRATER
HYDRA
ANTLIA
CENTAURUS
PYXIS
PUPPIS
MON
CANIS MAJOR
VELA
COLUMBA
CARINA
CRUX
LUPUS
CIR
MUSCA
VOLANS
PICTOR
DOR
DORADO
Canopus
Avior
Aspidiske
Miaplacides
Hadar
Acrux
Gacrux
Mimosa
Wezen
Adhara
Aludra
Omega Centauri
Jewel Box
Coal Sack
Eta Carinae Nebula
Southern Pleiades
Vela Supernova Remnant
Galactic Equator
Wil Tirion

## Variable stars

| Star | Con | RA h m | Dec ° ′ | Range | Type | Period (days) | Spectrum |
|---|---|---|---|---|---|---|---|
| S | Cru | 12 54.4 | –58 26 | 6.2–6.9 | Cep | 4.69 | F–G |
| R | Hya | 13 29.7 | –23 17 | 3.5–10.9 | M | 389.6 | M |
| V744 | Cen | 13 40.0 | –49 57 | 5.1–6.6 | SR | 90: | M |
| T | Cen | 13 41.8 | –33 36 | 5.5–9.0 | SR | 90.44 | K–M |
| V766 | Cen | 13 47.2 | –62 35 | 6.2–7.5 | SD? | — | G |
| μ | Cen | 13 49.6 | –42 28 | 2.9–3.5 | Irr | — | B |
| V716 | Cen | 14 13.7 | –54 38 | 6.0–6.5 | EB | 1.49 | B |
| R | Cen | 14 16.6 | –59 55 | 5.3–11.8 | M | 546.2 | M |
| GG | Lup | 15 32.2 | –23 53 | 5.5–6.0 | SR | — | M |
| R | Nor | 15 36.0 | –49 30 | 6.5–13.9 | M | 492.7 | M |
| T | Nor | 15 44.1 | –54 59 | 6.2–13.6 | M | 242.6 | M |

## Double stars

| Star | Con | RA h m | Dec ° ′ | PA ° | Sep ″ | Magnitudes | |
|---|---|---|---|---|---|---|---|
| α | Cru | 12 26.6 | –63 06 | 115 | 4.4 | 1.4 + 1.9 | Acrux |
| | | | | 202 | 90.1 | 4.9 | |
| γ | Cru | 12 31.2 | –57 07 | 31 | 110.6 | 1.6 + 6.7 | Gacrux |
| | | | | 82 | 155.2 | 9.5 | |
| γ | Cen | 12 41.5 | –48 58 | 347 | 1.0 | 2.9 + 2.9 | Binary, 84.5 years |
| ι | Cru | 12 45.6 | –60 59 | 22 | 26.9 | 4.7 + 9.5 | |
| μ | Cru | 12 54.6 | –57 11 | 17 | 34.9 | 4.3 + 5.3 | |
| 3 | Cen | 13 51.8 | –33 00 | 108 | 7.9 | 4.5 + 6.0 | |
| 4 | Cen | 13 53.2 | –31 56 | 185 | 14.9 | 4.7 + 8.4 | |
| β | Cen | 14 03.8 | –60 22 | 251 | 1.3 | 0.7 + 3.9 | Hadar |
| α | Cen | 14 39.6 | –60 50 | 222 | 14.1 | 0.0 + 1.2 | Rigil Kent; binary, 79.9 years |
| | | | | 211: | 7860 (131′) | 11.0 | Proxima Centauri |
| α | Cir | 14 42.5 | –64 59 | 232 | 15.7 | 3.2 + 8.6 | |
| 54 | Hya | 14 46.0 | –25 27 | 126 | 8.6 | 5.1 + 7.1 | |
| 59 | Hya | 14 58.7 | –27 39 | 80 | 0.3 | 6.3 + 6.6 | Binary, 339.3 years |
| π | Lup | 15 05.1 | –47 03 | 73 | 1.4 | 4.6 + 4.7 | |
| κ | Lup | 15 11.9 | –48 44 | 144 | 26.8 | 3.9 + 5.8 | |
| μ | Lup | 15 18.5 | –47 53 | 142 | 1.2 | 5.1 + 5.2 | |
| | | | | 130 | 23.7 | 7.2 | |
| γ | Cir | 15 23.4 | –59 19 | 20 | 0.7 | 5.1 + 5.5 | Binary, 180 years |
| 2 | Sco | 15 53.6 | –25 20 | 274 | 2.5 | 4.7 + 7.4 | |
| ξ | Lup | 15 56.9 | –33 58 | 49 | 10.4 | 5.3 + 5.8 | |

## Open clusters

| NGC/IC | Other | Con | RA h m | Dec ° ′ | Mag | Diam ′ | N* | |
|---|---|---|---|---|---|---|---|---|
| 4103 | | Cru | 12 06.7 | –61 15 | 7.4 | 7 | 45 | |
| 4349 | | Cru | 12 24.5 | –61 54 | 7.4 | 16 | 30 | |
| 4439 | | Cru | 12 28.4 | –60 06 | 8.4 | 4 | | |
| — | H 5 | Cru | 12 29.0 | –60 46 | 7.1 | 6 | | |
| 4463 | | Mus | 12 30.0 | –64 48 | 7.2 | 5 | 30 | |
| 4609 | | Cru | 12 42.3 | –62 58 | 6.9 | 5 | 40 | |
| 4755 | | Cru | 12 53.6 | –60 20 | 4.2 | 10 | 50 | Jewel Box, κ Crucis |
| 4815 | | Mus | 12 58.0 | –64 57 | 8.6 | 3 | 100 | |
| 5138 | | Cen | 13 27.3 | –59 01 | 7.6 | 8 | 40 | |
| 5281 | | Cen | 13 46.6 | –62 54 | 5.9 | 5 | 40 | |
| 5316 | | Cen | 13 53.9 | –61 52 | 6.0 | 14 | 80 | |
| 5460 | | Cen | 14 07.6 | –48 19 | 5.6 | 25 | 40 | |
| 5606 | | Cen | 14 27.8 | –59 38 | 7.7 | 3 | 15 | |
| 5617 | | Cen | 14 29.8 | –60 43 | 6.3 | 10 | 80 | |

## Open clusters *(continued)*

| NGC/IC | Other | Con | RA h m | Dec ° ′ | Mag | Diam ′ | N* |
|---|---|---|---|---|---|---|---|
| 5662 | | Cen | 14 35.2 | –56 33 | 5.5 | 12 | 70 |
| 5749 | | Lup | 14 48.9 | –54 31 | 8.8 | 8 | 30 |
| 5822 | | Lup | 15 05.2 | –54 21 | 6.5 | 40 | 150 |
| 5823 | | Cir | 15 05.7 | –55 36 | 7.9 | 10 | 100 |
| 5925 | | Nor | 15 27.7 | –54 31 | 8.4 | 15 | 120 |

## Globular clusters

| NGC/IC | Other | Con | RA h m | Dec ° ′ | Mag | Diam ′ | |
|---|---|---|---|---|---|---|---|
| 4590 | M68 | Hya | 12 39.5 | –26 45 | 8.2 | 12.0 | |
| 5139 | ω Cen | Cen | 13 26.8 | –47 29 | 3.7 | 36.3 | Omega Centauri |
| 5286 | | Cen | 13 46.4 | –51 22 | 7.6 | 9.1 | |
| 5824 | | Lup | 15 04.0 | –33 04 | 9.0 | 6.2 | |
| 5897 | | Lib | 15 17.4 | –21 01 | 8.6 | 12.6 | |
| 5927 | | Lup | 15 28.0 | –50 40 | 8.3 | 12.0 | |
| 5946 | | Nor | 15 35.5 | –50 40 | 9.6 | 7.1 | |
| 5986 | | Lup | 15 46.1 | –37 47 | 7.1 | 9.8 | |

## Nebulae

| NGC/IC | Other | Con | RA h m | Dec ° ′ | Type | Diam ′ | Mag* | |
|---|---|---|---|---|---|---|---|---|
| — | Coal Sack | Cru | 12 53: | –63: | — | 400 × 300 | — | Dark nebula |
| 5367 | | Cen | 13 57.7 | –39 59 | R | 4 × 3 | | |

## Planetary nebulae

| NGC/IC | Other | Con | RA h m | Dec ° ′ | Mag p | Diam ″ | Mag* |
|---|---|---|---|---|---|---|---|
| — | PK303+40.1 | Hya | 12 53.6 | –22 52 | 12.0 | 709 | 9.6 |
| 5307 | | Cen | 13 51.1 | –51 12 | 12.1 | 13 | |
| I.4406 | | Cen | 14 22.4 | –44 09 | 10.6 | 28 | 14.7 |
| 5873 | | Lup | 15 12.8 | –38 08 | 13.3 | 3 | 13.6 |
| 5882 | | Lup | 15 16.8 | –45 39 | 10.5 | 7 | 12.0 |

## Galaxies

| NGC/IC | Other | Con | RA h m | Dec ° ′ | Mag | Size ′ | Type | |
|---|---|---|---|---|---|---|---|---|
| 4105 | | Hya | 12 06.7 | –29 46 | 12.0 | 2.4 × 1.9 | E2 | |
| 4106 | | Hya | 12 06.8 | –29 46 | 11.4 | 1.9 × 1.5 | E0 | |
| I.3370 | | Cen | 12 27.6 | –39 20 | 11.1 | 2.8 × 2.4 | E2 | |
| 4696 | | Cen | 12 48.8 | –41 19 | 10.7 | 3.5 × 3.2 | E1 | |
| 4945 | | Cen | 13 05.4 | –49 28 | 9.5 | 20.0 × 4.4 | SBc | |
| 4976 | | Cen | 13 08.6 | –49 30 | 10.2 | 4.3 × 2.6 | E4 | |
| 5061 | | Hya | 13 18.1 | –26 50 | 11.7 | 2.6 × 2.3 | E2 | |
| 5068 | | Vir | 13 18.9 | –21 02 | 10.8 | 6.9 × 6.3 | SBc | |
| 5078 | | Hya | 13 19.8 | –27 24 | 12: | 3.2 × 1.7 | Sa | |
| 5085 | | Hya | 13 20.3 | –24 26 | 11.9 | 3.4 × 3.0 | Sb | |
| 5101 | | Hya | 13 21.8 | –27 26 | 11.7 | 5.5 × 4.9 | SBa | |
| 5102 | | Cen | 13 22.0 | –36 38 | 9.7 | 9.3 × 3.5 | S0 | |
| 5128 | | Cen | 13 25.5 | –43 01 | 7.0 | 18.2 × 14.5 | S0 | |
| I.4296 | | Cen | 13 36.6 | –33 58 | 11.6 | 2.7 × 2.7 | E0 | |
| 5236 | M83 | Hya | 13 37.0 | –29 52 | 8.2 | 11.2 × 10.2 | Sc | Southern Pinwheel |
| 5253 | | Cen | 13 39.9 | –31 39 | 10.6 | 4.0 × 1.7 | E5 | |
| 5483 | | Cen | 14 10.4 | –43 19 | 12.0 | 3.1 × 2.8 | Sc | |
| 5530 | | Lup | 14 18.5 | –43 24 | 11.9 | 4.1 × 2.2 | Sb | |
| 5643 | | Lup | 14 32.7 | –44 10 | 10.7 | 4.6 × 4.1 | SBc | |

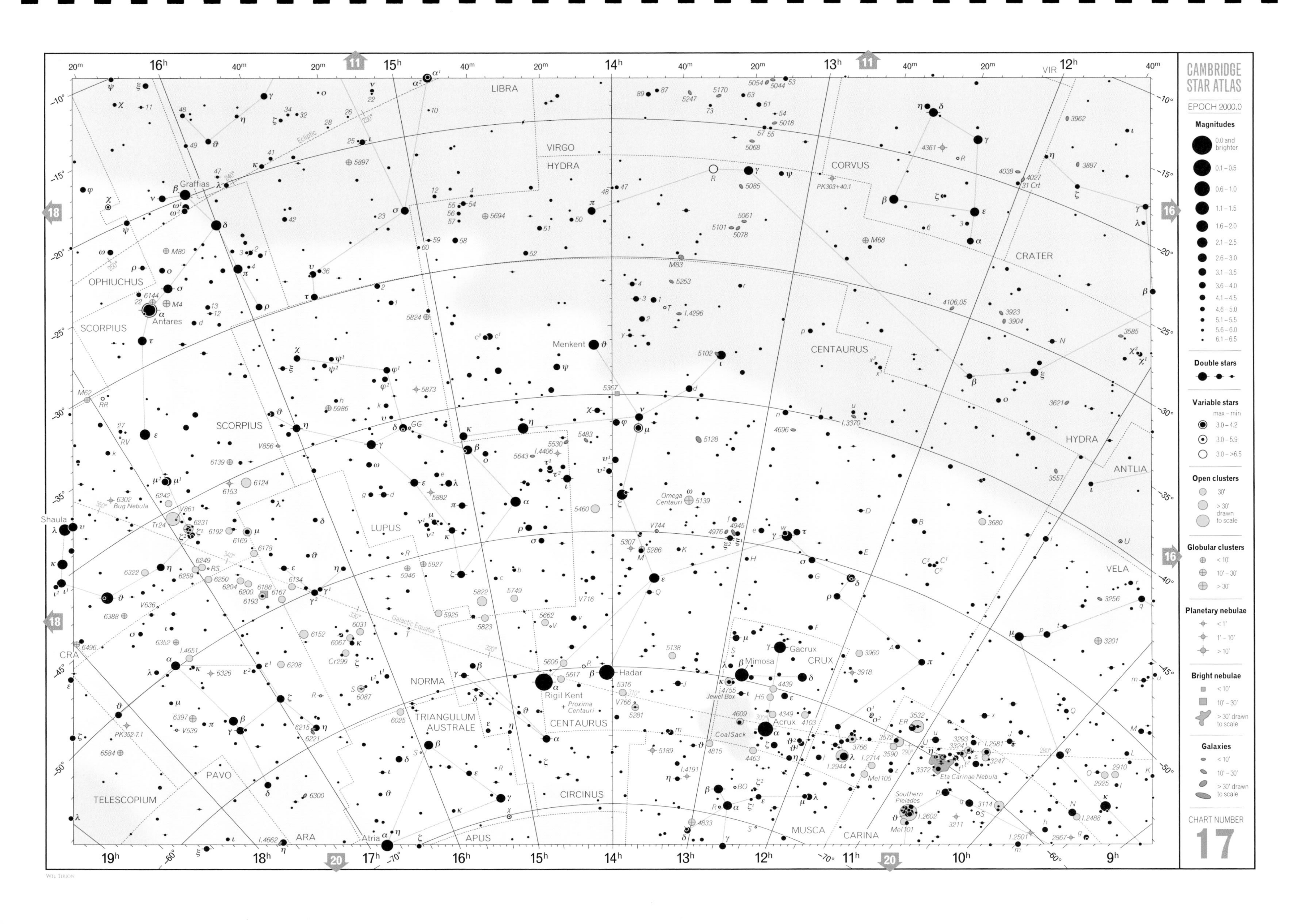

CAMBRIDGE STAR ATLAS
EPOCH 2000.0
Magnitudes
0.0 and brighter
0.1 – 0.5
0.6 – 1.0
1.1 – 1.5
1.6 – 2.0
2.1 – 2.5
2.6 – 3.0
3.1 – 3.5
3.6 – 4.0
4.1 – 4.5
4.6 – 5.0
5.1 – 5.5
5.6 – 6.0
6.1 – 6.5
Double stars
Variable stars
max – min
3.0 – 4.2
3.0 – 5.9
3.0 – >6.5
Open clusters
30'
> 30'
drawn to scale
Globular clusters
< 10'
10' – 30'
> 30'
Planetary nebulae
< 1'
1' – 10'
> 10'
Bright nebulae
< 10'
10' – 30'
> 30' drawn to scale
Galaxies
< 10'
10' – 30'
> 30' drawn to scale
CHART NUMBER
17
LIBRA
VIRGO
HYDRA
CORVUS
CRATER
CENTAURUS
ANTLIA
VELA
OPHIUCHUS
SCORPIUS
LUPUS
NORMA
TRIANGULUM AUSTRALE
CIRCINUS
CRUX
MUSCA
CARINA
APUS
ARA
PAVO
TELESCOPIUM
CRA
VIR
Graffias
Antares
Shaula
Menkent
Hadar
Rigil Kent
Proxima Centauri
Omega Centauri
Gacrux
Mimosa
Acrux
Jewel Box
Coal Sack
Eta Carinae Nebula
Southern Pleiades
Bug Nebula
Atria
Ecliptic
Galactic Equator

## Chart 18 *RA 16ʰ to 20ʰ, declination –20° to –65°*

### Variable stars

| Star | Con | RA h m | Dec ° ′ | Range | Type | Period (days) | Spectrum | |
|---|---|---|---|---|---|---|---|---|
| S | TrA | 16 01.2 | –63 47 | 6.1–6.8 | Cep | 6.32 | F | |
| S | Nor | 16 18.9 | –57 54 | 6.1–6.8 | Cep | 9.75 | F–G | |
| α | Sco | 16 29.4 | –26 26 | 0.9–1.8 | SR | 1,733 | M + B | Antares |
| R | Ara | 16 39.7 | –57 00 | 6.0–6.9 | EA | 4.43 | B | |
| RS | Sco | 16 55.6 | –45 06 | 6.2–13.0 | M | 320.1 | M | |
| RR | Sco | 16 56.6 | –30 35 | 5.0–12.4 | M | 279.40 | B | |
| V861 | Sco | 16 56.6 | –40 49 | 6.1–6.7 | EB | 7.85 | B | |
| RV | Sco | 16 58.3 | –33 37 | 6.6–7.5 | Cep | 6.06 | F–G | |
| V636 | Sco | 17 22.8 | –45 37 | 6.0–6.9 | Cep | 6.80 | G | |
| BM | Sco | 17 41.0 | –32 13 | 5.0–6.9 | SR | 850: | K | |
| X | Sgr | 17 47.6 | –27 50 | 4.2–4.8 | Cep | 7.01 | F | |
| V539 | Ara | 17 50.5 | –53 37 | 5.7–6.2 | EA | 3.17 | B + B | |
| W | Sgr | 18 05.0 | –29 35 | 4.3–5.1 | Cep | 7.59 | F–G | |
| RS | Sgr | 18 17.6 | –34 06 | 6.0–6.9 | EA | 2.42 | B | |
| V1017 | Sgr | 18 32.1 | –29 24 | 6.2–14.7 | ZA? | — | G | |
| λ | Pav | 18 52.2 | –62 11 | 3.4–4.3 | Irr | — | B | |
| RY | Sgr | 19 16.5 | –33 31 | 6.0–15.0 | RCB | — | G | |
| S | Pav | 19 55.2 | –59 12 | 6.6–10.4 | SR | 386.3 | M | |
| RR | Sgr | 19 55.9 | –29 11 | 5.6–14.0 | M | 334.6 | M | |
| RU | Sgr | 19 58.7 | –41 51 | 6.0–13.8 | M | 240.3 | M | |

### Double stars

| Star | Con | RA h m | Dec ° ′ | PA ° | Sep ″ | Magnitudes | |
|---|---|---|---|---|---|---|---|
| 12 | Sco | 16 12.3 | –28 25 | 73 | 4.0 | 5.9 + 7.9 | |
| σ | Sco | 16 21.2 | –25 36 | 273 | 20.0 | 2.9 + 8.5 | |
| ε | Nor | 16 27.2 | –47 33 | 335 | 22.8 | 4.8 + 7.5 | |
| α | Sco | 16 29.4 | –26 26 | 275 | 2.9 | 1.2 v + 5.4 | Antares; var.; bin., 878 y. |
| 36 | Oph | 17 15.3 | –26 36 | 146 | 4.9 | 5.1 + 5.1 | Binary, 548.7 years |
| γ | Ara | 17 25.4 | –56 23 | 328 | 17.9 | 3.3 + 10.3 | |
| η | Sgr | 18 17.6 | –36 46 | 105 | 3.6 | 3.2 + 7.8 | |
| ξ | Pav | 18 23.2 | –61 30 | 154 | 3.3 | 4.4 + 8.6 | |
| 21 | Sgr | 18 25.3 | –20 32 | 289 | 1.8 | 4.9 + 7.4 | |
| λ | CrA | 18 43.8 | –38 19 | 214 | 29.2 | 5.1 + 9.7 | |
| γ | CrA | 19 06.4 | –37 04 | 55 | 1.3 | 4.8 + 5.1 | Binary, 120.4 years |
| π | Sgr | 19 09.8 | –21 01 | 150 | 0.1 | 3.7 + 3.7 | |
| | | | | 122 | 0.4 | 5.9 | |
| β¹ | Sgr | 19 22.6 | –44 28 | 77 | 28.3 | 4.0 + 7.1 | |
| 52 | Sgr | 19 36.7 | –24 53 | 170 | 2.5 | 4.7 + 9.2 | |

### Open clusters

| NGC/IC | Other | Con | RA h m | Dec ° ′ | Mag | Diam ′ | N* | |
|---|---|---|---|---|---|---|---|---|
| 6025 | | TrA | 16 03.7 | –60 30 | 5.1 | 12 | 60 | |
| 6067 | | Nor | 16 13.2 | –54 13 | 5.6 | 13 | 100 | |
| — | Cr 299 | Cep | 16 18.4 | –55 07 | 6.9 | 20 | | |
| 6087 | | Nor | 16 18.9 | –57 54 | 5.4 | 12 | 40 | |
| 6124 | | Sco | 16 25.6 | –40 40 | 5.8 | 29 | 100 | |
| 6134 | | Nor | 16 27.7 | –49 09 | 7.2 | 7 | | |
| 6169 | | Lup | 16 34.1 | –44 03 | 6.6 | 7 | 40 | Surrounds μ Normae |
| 6167 | | Nor | 16 34.4 | –49 36 | 6.7 | 8 | | |
| 6178 | | Sco | 16 35.7 | –45 38 | 7.2 | 4 | 12 | |
| 6193 | | Ara | 16 41.3 | –48 46 | 5.2 | 15 | | |
| 6200 | | Ara | 16 44.2 | –47 29 | 7.4 | 12 | 40 | |
| 6208 | | Ara | 16 49.5 | –53 49 | 7.2 | 16 | 60 | |
| 6231 | | Sco | 16 54.0 | –41 48 | 2.6 | 15 | | |
| 6242 | | Sco | 16 55.6 | –39 30 | 6.4 | 9 | | |
| 6250 | | Ara | 16 58.0 | –45 48 | 5.9 | 8 | 60 | |
| 6322 | | Sco | 17 18.5 | –42 57 | 6.0 | 10 | 30 | |
| I.4651 | | Ara | 17 24.7 | –49 57 | 6.9 | 12 | 80 | |
| 6383 | | Sco | 17 34.8 | –32 34 | 5.5 | 5 | 40 | |
| 6405 | M6 | Sco | 17 40.1 | –32 13 | 4.2 | 15 | 80 | Butterfly Cluster |
| 6416 | | Sco | 17 44.4 | –32 21 | 4.7 | 18 | 40 | |
| 6425 | | Sgr | 17 46.9 | –31 32 | 7.2 | 8 | 35 | |
| 6475 | M7 | Sco | 17 53.9 | –34 49 | 3.3 | 80 | 80 | Ptolemy's Cluster |
| 6520 | | Sgr | 18 03.4 | –27 54 | 7.6 | 6 | 60 | |
| 6531 | M21 | Sgr | 18 04.6 | –22 30 | 5.9 | 13 | 17 | |
| 6530 | | Sgr | 18 04.8 | –24 20 | 4.6 | 15 | | In M8 |

### Globular clusters

| NGC/IC | Other | Con | RA h m | Dec ° ′ | Mag | Diam ′ |
|---|---|---|---|---|---|---|
| 6093 | M80 | Sco | 16 17.0 | –22 59 | 7.2 | 8.9 |
| 6121 | M4 | Sco | 16 23.6 | –26 32 | 5.9 | 26.3 |
| 6144 | | Sco | 16 27.3 | –26 02 | 9.1 | 9.3 |
| 6139 | | Sco | 16 27.7 | –38 51 | 9.2 | 5.5 |
| 6266 | M62 | Oph | 17 01.2 | –30 07 | 6.6 | 14.1 |
| 6273 | M19 | Oph | 17 02.6 | –26 16 | 7.2 | 13.5 |
| 6284 | | Oph | 17 04.5 | –24 46 | 9.0 | 5.6 |
| 6287 | | Oph | 17 05.2 | –22 42 | 9.2 | 5.1 |
| 6293 | | Oph | 17 10.2 | –26 35 | 8.2 | 7.9 |
| 6304 | | Oph | 17 14.5 | –29 28 | 8.4 | 6.8 |
| 6316 | | Oph | 17 16.6 | –28 08 | 9.0 | 4.9 |
| 6352 | | Ara | 17 25.5 | –48 25 | 8.2 | 7.1 |
| 6388 | | Sco | 17 36.3 | –44 44 | 6.9 | 8.7 |
| 6397 | | Ara | 17 40.7 | –53 40 | 5.7 | 25.7 |
| 6441 | | Sco | 17 50.2 | –37 03 | 7.4 | 7.8 |
| 6496 | | Sco | 17 59.0 | –44 16 | 9.2 | 6.9 |
| 6544 | | Sgr | 18 07.3 | –25 00 | 8.3 | 8.9 |
| 6541 | | CrA | 18 08.0 | –43 42 | 6.7 | 13.1 |
| 6553 | | Sgr | 18 09.3 | –25 54 | 8.3 | 8.1 |
| 6569 | | Sgr | 18 13.6 | –31 50 | 8.7 | 5.8 |
| 6584 | | CrA | 18 18.6 | –52 13 | 9.2 | 7.9 |
| 6624 | | Sgr | 18 23.7 | –30 22 | 8.3 | 5.9 |
| 6626 | M28 | Sgr | 18 24.5 | –24 52 | 6.9 | 11.2 |
| 6638 | | Sgr | 18 30.9 | –25 30 | 9.2 | 5.0 |
| 6637 | M69 | Sgr | 18 31.4 | –32 21 | 7.7 | 7.1 |
| 6642 | | Sgr | 18 31.9 | –23 29 | 8.8 | 4.5 |
| 6652 | | Sgr | 18 35.8 | –32 59 | 8.9 | 3.5 |
| 6656 | M22 | Sgr | 18 36.4 | –23 54 | 5.1 | 24.0 |
| 6681 | M70 | Sgr | 18 43.2 | –32 18 | 8.1 | 7.8 |
| 6715 | M54 | Sgr | 18 55.1 | –30 29 | 7.7 | 9.1 |
| 6717 | | Sgr | 18 55.1 | –22 42 | 9.2 | 3.9 |
| 6723 | | Sgr | 18 59.6 | –36 38 | 7.3 | 11.0 |
| 6752 | | Pav | 19 10.9 | –59 59 | 5.4 | 20.4 |
| 6809 | M55 | Sgr | 19 40.0 | –30 58 | 7.0 | 19.0 |

### Bright diffuse nebulae

| NGC/IC | Other | Con | RA h m | Dec ° ′ | Type | Diam ′ | Mag* | |
|---|---|---|---|---|---|---|---|---|
| 6188, 6193 | | Ara | 16 40.5 | –48 47 | E + R | 20 × 12 | | |
| 6514 | M20 | Sgr | 18 02.6 | –23 02 | E + R | 29 × 27 | 7.6 | Trifid Nebula |
| 6523 | M8 | Sgr | 18 03.8 | –24 23 | E | 90 × 40 | 0.0 | Lagoon Nebula |

### Planetary nebulae

| NGC/IC | Other | Con | RA h m | Dec ° ′ | Mag p | Diam ″ | Mag* | |
|---|---|---|---|---|---|---|---|---|
| 6153 | | Sco | 16 31.5 | –40 15 | 11.5 | 25 | | |
| I.4634 | | Oph | 17 01.6 | –21 50 | 10.7 | 9 | 15 p | |
| 6302 | | Sco | 17 13.7 | –37 06 | 12.8 | 50 | | Bug Nebula |
| 6326 | | Ara | 17 20.8 | –51 45 | 12.2 | 14 | 13.5 | |
| 6369 | | Oph | 17 29.3 | –23 46 | 12.9 | 30/66 | 14.7 | |
| — | PK352–7.1 | Ara | 18 00.2 | –38 50 | 11.4 | 25 | | |
| I.4699 | | CrA | 18 18.5 | –45 59 | 11.9 | 10 | | |
| 6629 | | Sgr | 18 25.7 | –23 12 | 11.6 | 15 | 12.8 | |
| — | PK3–14.1 | Sgr | 18 55.6 | –32 16 | 10.9 | 4 | 18 p | |
| I.1297 | | CrA | 19 17.4 | –39 37 | 10.7 | 7 | 12.9 | |

### Galaxies

| NGC/IC | Other | Con | RA h m | Dec ° ′ | Mag | Size ′ | Type |
|---|---|---|---|---|---|---|---|
| 6215 | | Ara | 16 51.1 | –58 59 | 11.8 | 2.0 × 1.6 | Sc |
| 6221 | | Ara | 16 52.8 | –59 13 | 11.5 | 3.2 × 2.3 | SBc |
| 6300 | | Ara | 17 17.0 | –62 49 | 11.1 | 5.4 × 3.5 | SBc |
| I.4662 | | Pav | 17 47.1 | –64 38 | 11.4 | 2.2 × 1.4 | Irr |
| 6744 | | Pav | 19 09.8 | –63 51 | 9.0 | 15.5 × 10.2 | SBb |
| 6753 | | Pav | 19 11.4 | –57 03 | 11.9 | 2.5 × 2.2 | Sb |

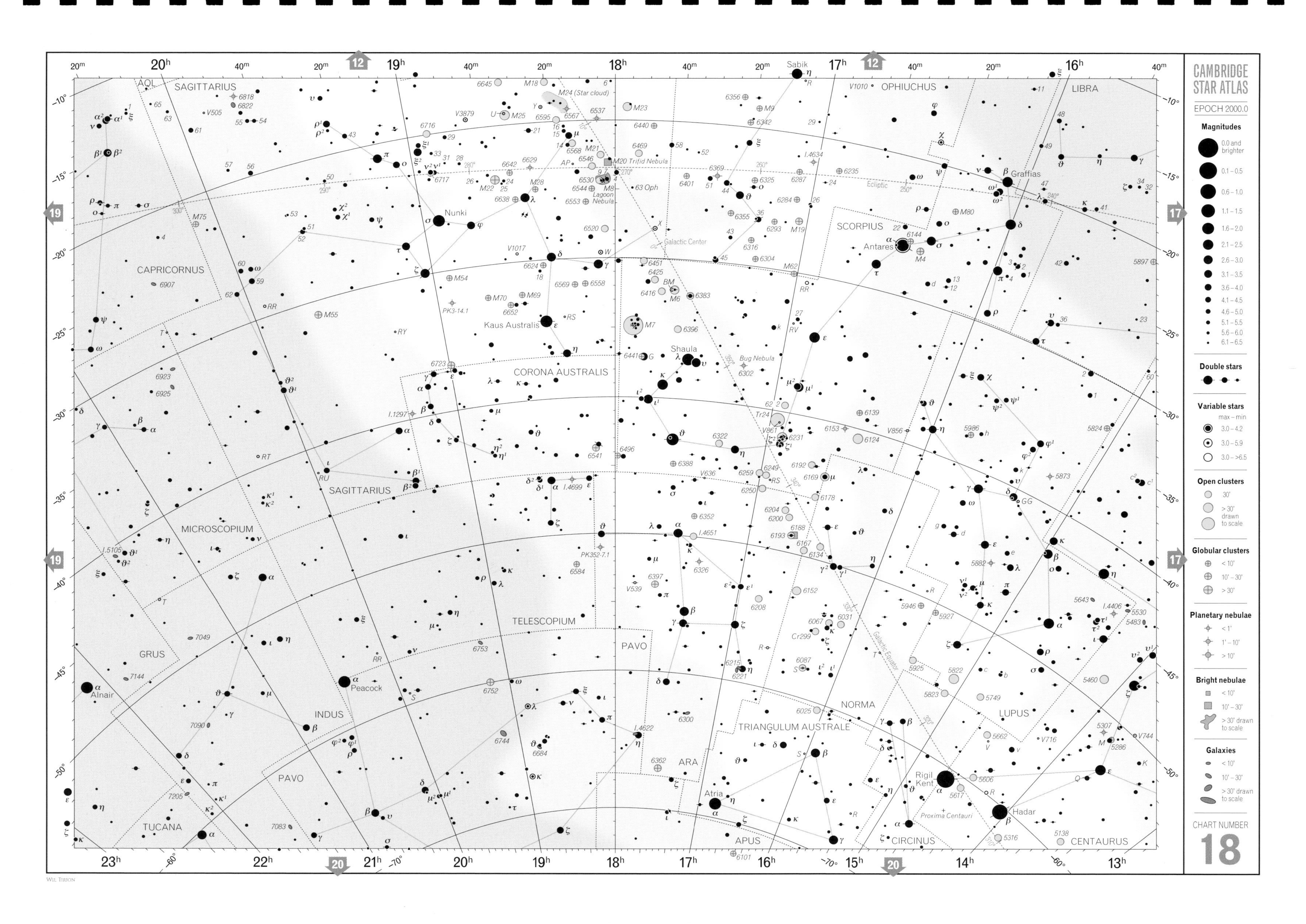

CAMBRIDGE STAR ATLAS
EPOCH 2000.0
Magnitudes
0.0 and brighter
0.1 – 0.5
0.6 – 1.0
1.1 – 1.5
1.6 – 2.0
2.1 – 2.5
2.6 – 3.0
3.1 – 3.5
3.6 – 4.0
4.1 – 4.5
4.6 – 5.0
5.1 – 5.5
5.6 – 6.0
6.1 – 6.5
Double stars
Variable stars
max – min
3.0 – 4.2
3.0 – 5.9
3.0 – >6.5
Open clusters
30'
> 30' drawn to scale
Globular clusters
< 10'
10' – 30'
> 30'
Planetary nebulae
< 1'
1' – 10'
> 10'
Bright nebulae
< 10'
10' – 30'
> 30' drawn to scale
Galaxies
< 10'
10' – 30'
> 30' drawn to scale
CHART NUMBER
18
SAGITTARIUS
AQL
CAPRICORNUS
MICROSCOPIUM
GRUS
INDUS
PAVO
TUCANA
CORONA AUSTRALIS
TELESCOPIUM
ARA
TRIANGULUM AUSTRALE
NORMA
LUPUS
APUS
CIRCINUS
CENTAURUS
SCORPIUS
OPHIUCHUS
LIBRA
Nunki
Kaus Australis
Shaula
Antares
Graffias
Sabik
Peacock
Alnair
Atria
Rigil Kent
Hadar
Proxima Centauri
M24 (Star cloud)
M20 Trifid Nebula
M8 Lagoon Nebula
Bug Nebula
Galactic Center
Galactic Equator
Ecliptic
63 Oph
Wil Tirion

## Variable stars

| Star | Con | RA h m | Dec ° ′ | Range | Type | Period (days) | Spectrum |
|---|---|---|---|---|---|---|---|
| RR | Tel | 20 04.2 | –55 43 | 6.5–16.5 p | ZA | — | F |
| RT | Sgr | 20 17.7 | –39 07 | 6.0–14.1 | M | 305.3 | M |
| T | Mic | 20 27.9 | –28 16 | 7.7–9.6 p | SR | 344 | M |
| T | Ind | 21 20.2 | –45 01 | 5.5–6.5 | SR | 320: | M |
| T | Gru | 22 25.7 | –37 34 | 7.8–12.3 p | M | 136.5 | M |
| S | Gru | 22 26.1 | –48 26 | 6.0–15.0 | M | 401.4 | M |

## Double stars

| Star | Con | RA h m | Dec ° ′ | PA ° | Sep ″ | Magnitudes |
|---|---|---|---|---|---|---|
| α | Mic | 20 50.0 | –33 47 | 166 | 20.5 | 5.0 + 10.0 |
| ϑ | Ind | 21 19.9 | –53 27 | 275 | 6.0 | 4.5 + 7.0 |
| η | PsA | 22 00.8 | –28 27 | 115 | 1.7 | 5.8 + 6.8 |
| 41 | Aqr | 22 14.3 | –21 04 | 114 | 5.0 | 5.6 + 7.1 |
| δ | Tuc | 22 27.3 | –64 58 | 282 | 6.9 | 4.5 + 9.0 |
| β | PsA | 22 31.5 | –32 21 | 172 | 30.3 | 4.9 + 7.9 |
| γ | PsA | 22 52.5 | –32 53 | 262 | 4.2 | 4.5 + 8.0 |
| δ | PsA | 22 55.9 | –32 32 | 244 | 5.0 | 4.2 + 9.2 |
| υ | Gru | 23 06.9 | –38 54 | 211 | 1.1 | 5.7 + 8.0 |
| ϑ | Gru | 23 06.9 | –43 31 | 75 | 1.1 | 4.5 + 7.0 |

## Globular clusters

| NGC/IC | Other | Con | RA h m | Dec ° ′ | Mag | Diam ′ |
|---|---|---|---|---|---|---|
| 6864 | M75 | Sgr | 20 06.1 | –21 55 | 8.6 | 6.0 |
| 7099 | M30 | Cap | 21 40.4 | –23 11 | 7.5 | 11.0 |

## Planetary nebula

| NGC/IC | Other | Con | RA h m | Dec ° ′ | Mag p | Diam ″ | Mag* | |
|---|---|---|---|---|---|---|---|---|
| 7293 | | Aqr | 22 29.6 | –20 48 | 7.3 | 769 | 13.8 | Helix Nebula |

## Galaxies

| NGC/IC | Other | Con | RA h m | Dec ° ′ | Mag | Size ′ | Type |
|---|---|---|---|---|---|---|---|
| 6907 | | Cap | 20 25.1 | –24 49 | 11.3 | 3.4 × 3.0 | SBb |
| 6923 | | Mic | 20 31.7 | –30 50 | 12.1 | 2.5 × 1.4 | Sb |
| 6925 | | Mic | 20 34.3 | –31 59 | 11.3 | 4.1 × 1.6 | Sb |
| 7049 | | Ind | 21 19.0 | –48 34 | 10.7 | 2.8 × 2.2 | S0 |
| I.5105 | | Mic | 21 24.4 | –40 37 | 11.5 | 2.5 × 1.5 | E4 |
| 7083 | | Ind | 21 35.7 | –63 54 | 11.8 | 4.5 × 2.9 | Sb |
| 7090 | | Ind | 21 36.5 | –54 33 | 11.1 | 7.1 × 1.4 | SBc |
| 7144 | | Gru | 21 52.7 | –48 15 | 10.7 | 3.5 × 3.5 | E0 |
| 7172 | | PsA | 22 02.0 | –31 52 | 11.9 | 2.2 × 1.3 | S |
| 7174 | | PsA | 22 02.1 | –31 59 | 12.6 | 1.3 × 0.7 | S |
| 7176 | | PsA | 22 02.1 | –31 59 | 11.9 | 1.3 × 1.3 | E0 |
| 7184 | | Aqr | 22 02.7 | –20 49 | 12.0 | 5.8 × 1.8 | Sb |
| 7205 | | Ind | 22 08.5 | –57 25 | 11.4 | 4.3 × 2.2 | Sb |
| 7221 | | PsA | 22 11.3 | –30 37 | 12 : | 2.2 × 1.9 | SBb |
| 7314 | | PsA | 22 35.8 | –26 03 | 10.9 | 4.6 × 2.3 | Sc |
| 7410 | | Gru | 22 55.0 | –39 40 | 10.4 | 5.5 × 2.0 | SBa |
| 7412 | | Gru | 22 55.8 | –42 39 | 11.4 | 4.0 × 3.1 | SBb |
| 7418 | | Gru | 22 56.6 | –37 02 | 11.4 | 3.3 × 2.8 | SBc |
| I.1459 | | Gru | 22 57.2 | –36 28 | 10.0 | 3.4 × 2.5 | E3 |
| I.5267 | | Gru | 22 57.2 | –43 24 | 10.5 | 5.0 × 4.1 | S0 |
| 7424 | | Gru | 22 57.3 | –41 04 | 10.5 | 7.6 × 6.8 | SBc |
| 7456 | | Gru | 23 02.1 | –39 35 | 11.9 | 5.9 × 1.8 | Sc |
| 7496 | | Gru | 23 09.8 | –43 26 | 11.1 | 3.5 × 2.8 | SBb |
| 7531 | | Gru | 23 14.8 | –43 36 | 11.3 | 3.5 × 1.5 | Sb |
| 7552 | | Gru | 23 16.2 | –42 35 | 10.7 | 3.5 × 2.5 | SBb |
| 7582 | | Gru | 23 18.4 | –42 22 | 10.6 | 4.6 × 2.2 | SBb |
| 7599 | | Gru | 23 19.3 | –42 15 | 11.4 | 4.4 × 1.5 | Sc |
| I.5332 | | Scl | 23 34.5 | –36 06 | 10.6 | 6.6 × 5.1 | Sd |
| 7713 | | Scl | 23 36.5 | –37 56 | 11.6 | 4.3 × 2.0 | SBd |
| 7755 | | Scl | 23 47.9 | –30 31 | 11.8 | 3.7 × 3.0 | SBb |
| 7793 | | Scl | 23 57.8 | –32 35 | 9.1 | 9.1 × 6.6 | Sd |

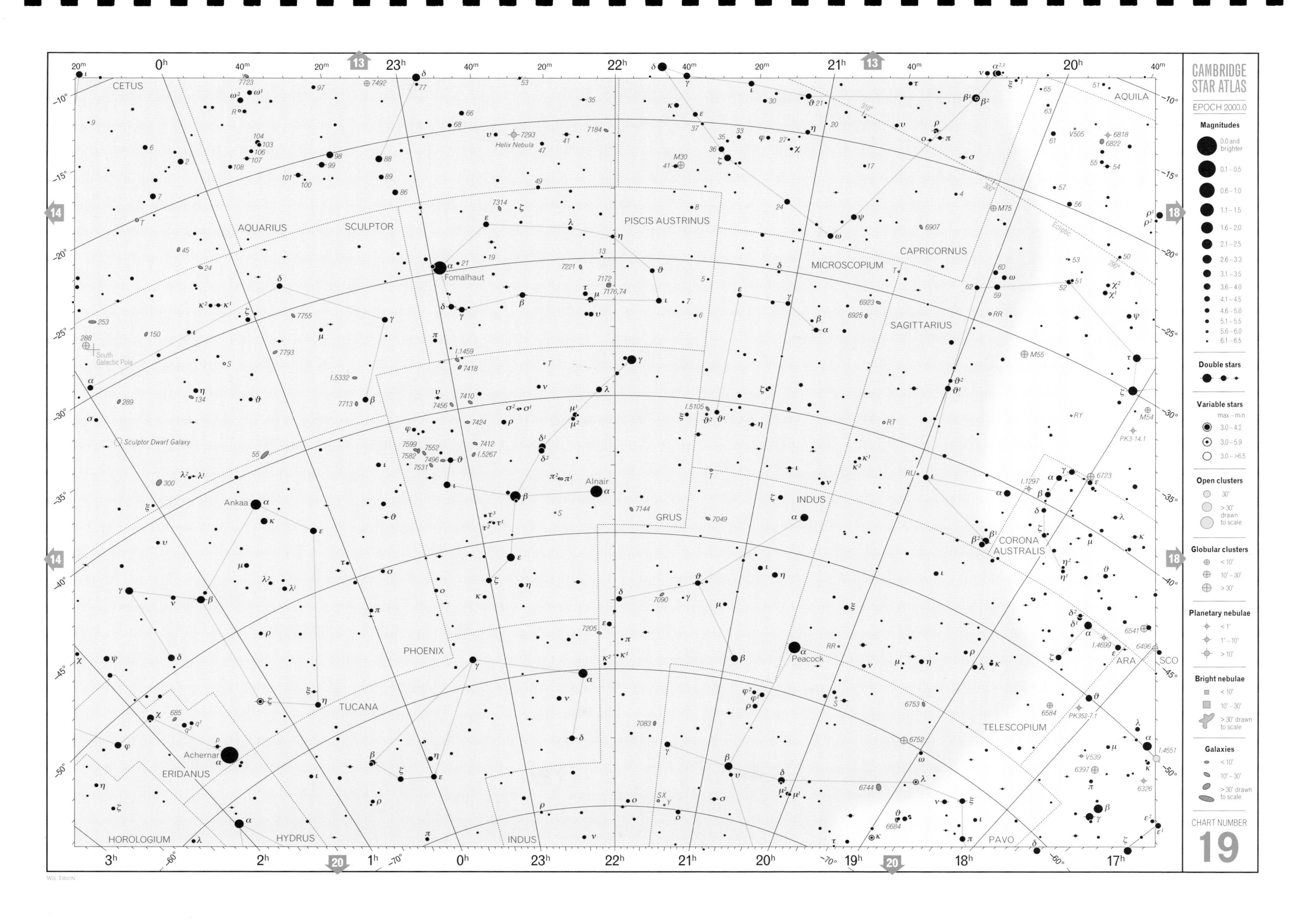
CAMBRIDGE STAR ATLAS
EPOCH 2000.0
Magnitudes
0.0 and brighter
0.1 – 0.5
0.6 – 1.0
1.1 – 1.5
1.6 – 2.0
2.1 – 2.5
2.6 – 3.0
3.1 – 3.5
3.6 – 4.0
4.1 – 4.5
4.6 – 5.0
5.1 – 5.5
5.6 – 6.0
6.1 – 6.5
Double stars
Variable stars
max – min
3.0 – 4.2
3.0 – 5.9
3.0 – >6.5
Open clusters
30'
> 30' drawn to scale
Globular clusters
< 10'
10' – 30'
> 30'
Planetary nebulae
< 1'
1' – 10'
> 10'
Bright nebulae
< 10'
10' – 30'
> 30' drawn to scale
Galaxies
< 10'
10' – 30'
> 30' drawn to scale
CHART NUMBER
19
CETUS
AQUARIUS
SCULPTOR
PISCIS AUSTRINUS
MICROSCOPIUM
CAPRICORNUS
SAGITTARIUS
AQUILA
CORONA AUSTRALIS
INDUS
GRUS
PHOENIX
TUCANA
ERIDANUS
HOROLOGIUM
HYDRUS
PAVO
TELESCOPIUM
ARA
SCO
Fomalhaut
Alnair
Ankaa
Achernar
Peacock
Helix Nebula
Sculptor Dwarf Galaxy
South Galactic Pole
Ecliptic
Wil Tirion

## Variable stars

| Star | Con | RA h m | Dec ° ′ | Range | Type | Period (days) | Spectrum |
|---|---|---|---|---|---|---|---|
| U | Men | 04 09.6 | −81 51 | 8.0–10.9 p | M | 407 | M |
| R | Oct | 05 26.1 | −86 23 | 6.4–13.2 | M | 405.6 | M |
| TZ | Men | 05 30.2 | −84 47 | 6.2–6.9 | EA | 8.57 | B |
| RS | Cha | 08 43.2 | −79 04 | 6.0–6.7 | EA + δSct | 1.67 | A + A |
| S | Mus | 12 12.8 | −70 09 | 5.9–6.4 | Cep | 9.66 | F |
| BO | Mus | 12 34.9 | −67 45 | 6.0–6.7 | Irr | — | M |
| R | Mus | 12 42.1 | −69 24 | 5.9–6.7 | Cep | 7.48 | B |
| J | Aps | 13 08.1 | −65 18 | 6.4–8.6 | SR | 199 | B |
| X | TrA | 15 14.3 | −70 05 | 5.0–6.4 | Irr | — | K |
| R | TrA | 15 19.8 | −66 30 | 6.4–6.9 | Cep | 3.39 | G |
| κ | Pav | 18 16.9 | −67 14 | 3.9–4.6 | Cep | 9.09 | F |
| Y | Pav | 21 24.3 | −69 44 | 5.6–7.3 | SR | 233.3 | M |
| SX | Pav | 21 28.7 | −69 30 | 5.4–6.0 | SR | 50: | M |

## Double stars

| Star | Con | RA h m | Dec ° ′ | PA ° | Sep ″ | Magnitudes | |
|---|---|---|---|---|---|---|---|
| κ | Tuc | 01 15.8 | −68 53 | 336 | 5.4 | 5.1 + 7.3 | |
| γ | Vol | 07 08.8 | −70 30 | 300 | 13.6 | 4.0 + 5.9 | |
| ζ | Vol | 07 41.8 | −72 36 | 116 | 16.7 | 4.0 + 9.8 | |
| κ | Vol | 08 19.8 | −71 31 | 57 | 65.0 | 5.4 + 5.7 | |
| ε | Cha | 11 59.6 | −78 13 | 188 | 0.9 | 5.4 + 6.0 | |
| β | Mus | 12 46.3 | −68 06 | 43 | 1.3 | 3.7 + 4.0 | Binary, 383.1 years |
| ι | Oct | 12 55.0 | −85 07 | 230 | 0.6 | 6.0 + 6.5 | |
| ϑ | Mus | 13 08.1 | −65 18 | 187 | 5.3 | 5.7 + 7.3 | |
| $\delta^{1,2}$ | Aps | 16 20.3 | −78 42 | 12 | 102.9 | 4.7 + 5.1 | |
| $\mu^2$ | Oct | 20 41.7 | −75 21 | 17 | 17.4 | 7.1 + 7.6 | |
| λ | Oct | 21 50.9 | −82 43 | 70 | 3.1 | 5.4 + 7.7 | |

## Open cluster

| NGC/IC | Other | Con | RA h m | Dec ° ′ | Mag | Diam ′ | N* | |
|---|---|---|---|---|---|---|---|---|
| — | Mel 101 | Car | 10 42.1 | −65 06 | 8.0: | 14 | 50 | Close to I.2602 |

## Globular clusters

| NGC/IC | Other | Con | RA h m | Dec ° ′ | Mag | Diam ′ | |
|---|---|---|---|---|---|---|---|
| 104 | | Tuc | 00 24.1 | −72 05 | 4.0 | 30.9 | 47 Tucanae |
| 362 | | Tuc | 01 03.2 | −70 51 | 6.6 | 12.9 | |
| 4372 | | Mus | 12 25.8 | −72 40 | 7.8 | 16.8 | |
| 4833 | | Mus | 12 59.6 | −70 53 | 7.4 | 13.5 | |
| 6101 | | Aps | 16 25.8 | −72 12 | 9.3 | 10.7 | |
| 6362 | | Ara | 17 31.9 | −67 03 | 8.3 | 10.7 | |

## Bright diffuse nebulae

| NGC/IC | Other | Con | RA h m | Dec ° ′ | Type | Diam ′ | Mag* | |
|---|---|---|---|---|---|---|---|---|
| 2070 | 30 Dor | Dor | 05 38.7 | −69 06 | E | 40 × 25 | | Tarantula Nebula in LMC[a] |

## Planetary nebulae

| NGC/IC | Other | Con | RA h m | Dec ° ′ | Mag p | Diam ″ | Mag* |
|---|---|---|---|---|---|---|---|
| I.2448 | | Car | 09 07.1 | −69 57 | 11.8 | 8 | 12.9 |
| I.4191 | | Mus | 13 08.8 | −67 39 | 12.0 | 5 | |
| 5189 | | Mus | 13 33.5 | −65 59 | 10.3 | 153 | 14.0 |

## Galaxies

| NGC/IC | Other | Con | RA h m | Dec ° ′ | Mag | Size ′ | Type |
|---|---|---|---|---|---|---|---|
| 1313 | | Ret | 03 18.3 | −66 30 | 9.4 | 8.5 × 6.6 | SB |
| 2442 | | Vol | 07 36.4 | −69 32 | 11.2 | 6.0 × 5.5 | SB |
| 3059 | | Car | 09 50.2 | −73 55 | 12.0 | 3.2 × 3.0 | SB |
| 6684 | | Pav | 18 49.0 | −65 11 | 11.4 | 3.7 × 2.7 | SB |

[a]LMC Large Magellanic Cloud: see page 81.

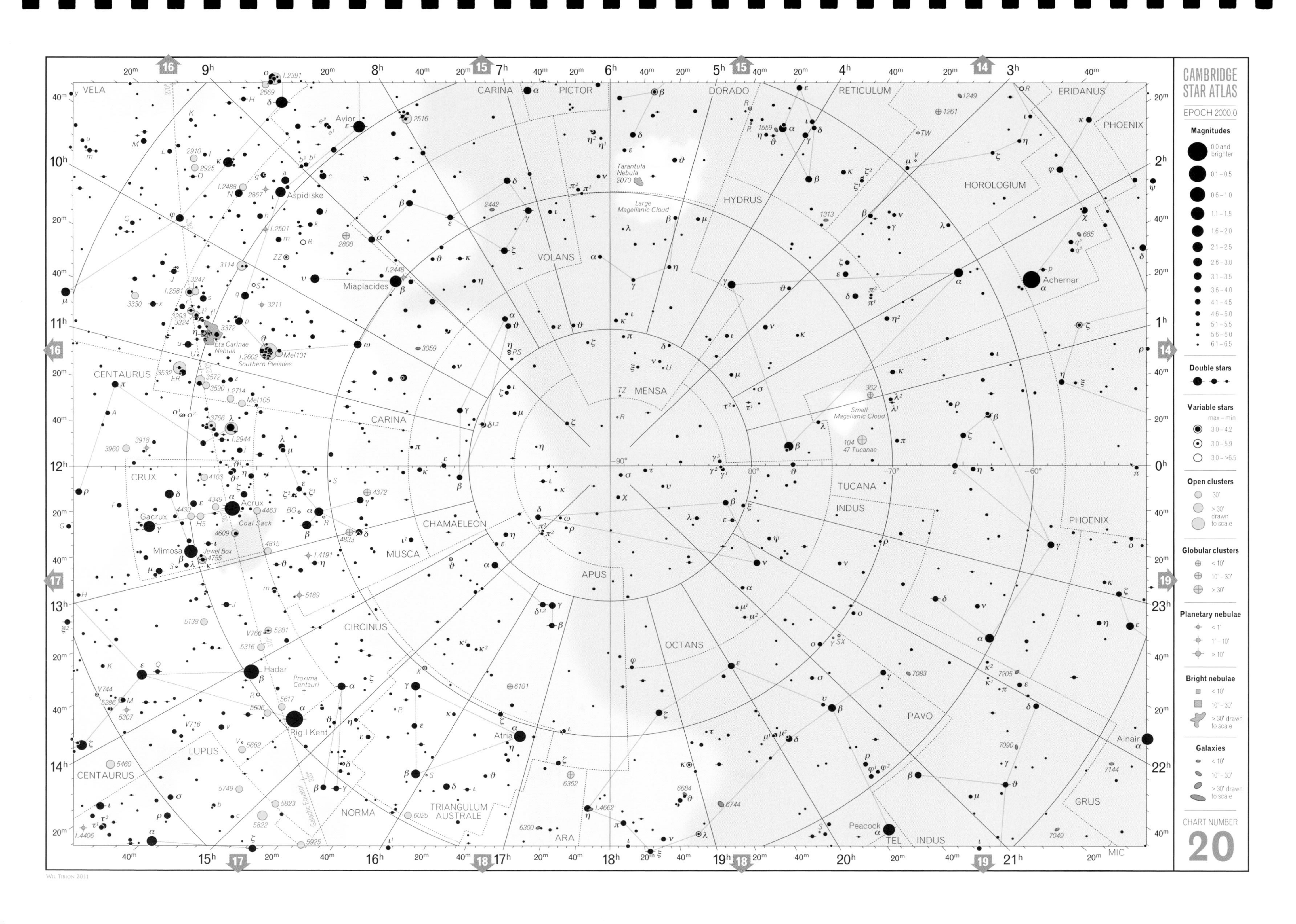
CAMBRIDGE STAR ATLAS
EPOCH 2000.0
Magnitudes
0.0 and brighter
0.1 – 0.5
0.6 – 1.0
1.1 – 1.5
1.6 – 2.0
2.1 – 2.5
2.6 – 3.0
3.1 – 3.5
3.6 – 4.0
4.1 – 4.5
4.6 – 5.0
5.1 – 5.5
5.6 – 6.0
6.1 – 6.5
Double stars
Variable stars
max – min
3.0 – 4.2
3.0 – 5.9
3.0 – >6.5
Open clusters
30'
> 30' drawn to scale
Globular clusters
< 10'
10' – 30'
> 30'
Planetary nebulae
< 1'
1' – 10'
> 10'
Bright nebulae
< 10'
10' – 30'
> 30' drawn to scale
Galaxies
< 10'
10' – 30'
> 30' drawn to scale
CHART NUMBER
20
VELA
CARINA
PICTOR
DORADO
RETICULUM
ERIDANUS
PHOENIX
HOROLOGIUM
HYDRUS
VOLANS
MENSA
TUCANA
INDUS
CENTAURUS
CRUX
CHAMAELEON
MUSCA
APUS
OCTANS
CIRCINUS
PAVO
LUPUS
NORMA
TRIANGULUM AUSTRALE
ARA
TEL
GRUS
MIC
Avior
Aspidiske
Miaplacides
Achernar
Eta Carinae Nebula
Southern Pleiades
Tarantula Nebula
Large Magellanic Cloud
Small Magellanic Cloud
47 Tucanae
Acrux
Gacrux
Mimosa
Jewel Box
Coal Sack
Hadar
Proxima Centauri
Rigil Kent
Atria
Peacock
Alnair
Galactic Equator
Wil Tirion 2011

# ALL-SKY MAPS

This part of *The Cambridge Star Atlas* consists of six all-sky maps to show the general distribution of different objects in the sky. Each of the six maps shows the whole sky in a so-called 'equal-area' projection; the *Mollweide* projection, named after the man who invented it.

It's obvious that you cannot show the inside of the whole celestial sphere on a simple flat map, without severe distortion. There are many different projections that can be used to create an all-sky map (or a map showing the whole surface of the Earth), and every projection has its advantages and disadvantages. Mollweide's is an 'equal-area' projection, meaning that in spite of the inevitable distortion the actual area covered by one square degree remains the same, no matter where on the map it is measured. So the distribution, or density of objects, is not influenced by the map's projection. Since that is the goal of these maps, the disadvantage, that the shapes of the constellations near the edges are extremely distorted, is taken for granted. If the constellation lines were not drawn, it would be hard to recognize them!

Unlike the main star charts, these maps are presented in galactic coordinates. The Galactic Equator is the central, horizontal line, marked 0°. On the main star charts it is represented as the blue dashed line, making an angle of almost 63° with the Celestial Equator. The Galactic Center (galactic longitude 0°, see star chart 18) lies close to the point where the constellations Sagittarius, Ophiuchus, and Scorpius meet, and is in the center of the galactic maps in this chapter. At the top and bottom of the maps the North Galactic Pole (NGP) and the South Galactic Pole (SGP) are marked. They can be found on the star charts 5 (in Coma Berenices) and 14 (in Sculptor) respectively.

## The constellations

The first map, on page 82, gives the positions of the constellations against this unusual grid. Stars down to magnitude 4.5 are plotted, plus some fainter stars to complete the constellation patterns, as on the seasonal sky maps.

## Distribution of open clusters

The second map, on page 83, shows the open star clusters as yellow disks with green outlines, plotted against the background of stars and constellations shown here blue against a black background. As explained in the introduction to the main star charts the open clusters are found near the plane of the Milky Way, so on this map you will find most of these objects close to the Galactic Equator. The map shows all clusters that are plotted on the star charts. This is also the case with the following maps of the globular clusters, the planetary nebulae, and the galaxies.

## Distribution of globular clusters

The third map, on page 84, shows the globular star clusters, using the same yellow symbols that are used on the star charts. You can see that the distribution is very different from the open clusters. The globular clusters form a huge halo around the Milky Way center, so they are much more scattered than the open clusters. But since we are not in the center of our galaxy (in fact, we are closer to the edge than to the center) we see most globular clusters in the direction of the Galactic Center (i.e., in the center of the map). In the opposite direction (near 180° galactic longitude, the left- and right-hand edges of the map) they are almost absent.

## Distribution of diffuse nebulae

The distribution of diffuse nebulae is shown on the fourth map, on page 85. Because on the star charts of *The Cambridge Star Atlas* only a limited number of bright diffuse nebulae are drawn, additional nebulae have been added to the all-sky map, to show their distribution properly. The light green squares are the ones drawn in *The Cambridge Star*

*Atlas*, while the darker green squares represent the additional nebulae, taken from *Sky Atlas 2000.0* (second edition). As you can see, the distribution of the nebulae is quite similar to that of the open star clusters. It brings us back to the disk of the Galaxy. Like the open clusters, the diffuse nebulae are found along the spiral arms of the Milky Way. In some areas you see gaps, where the nebulae are almost absent, but this is only an illusion. Large areas of dark nebulae and clouds of dust block the light of many remote bright nebulae as well as those of other objects.

## Distribution of planetary nebula

The fifth map, on page 86, shows the planetary nebulae (drawn as soft green disks with four spikes). Their distribution does not show any similarity with that of the open or the globular clusters. They are not exclusively found in the spiral arms of the Milky Way, like the open star clusters, and not in a halo, like the globular clusters. They fill a disk-like area that is much thicker than the main Milky Way disk, in which the open clusters and the diffuse nebulae are found.

## Distribution of galaxies

The sixth and final map shows where the brightest galaxies, plotted on the atlas charts, can be found. Since the galaxies, shown as red ovals, are not part of the Milky Way, but are in fact very distant 'Milky Ways' themselves, there is no relation between their distribution in the sky and the distribution of objects that belong to our own Milky Way. But, looking at the map, you will get the impression that there is a relationship, since other galaxies are almost absent in the area of the Milky Way close to the Galactic Equator. This, however, has nothing to do with their real distribution in space. The enormous amounts of gas and dust in our galaxy block the light of most of the distant galaxies seen in that direction. So they only appear to be absent. However, looking at the map, it is also obvious that the galaxies are not equally spread in the areas away from the Milky Way. They are grouped into what astronomers call 'clusters' and 'superclusters', with dimensions beyond our imagination.

In the bottom-right quadrant of the map you see two large galaxies. These are the Large and the Small Magellanic Clouds. The two clouds are regarded as satellites of our own Milky Way, and therefore are shown in white on the star charts (14, 15, and 20). However, since they are officially cataloged as galaxies, they have been plotted on this map as well.

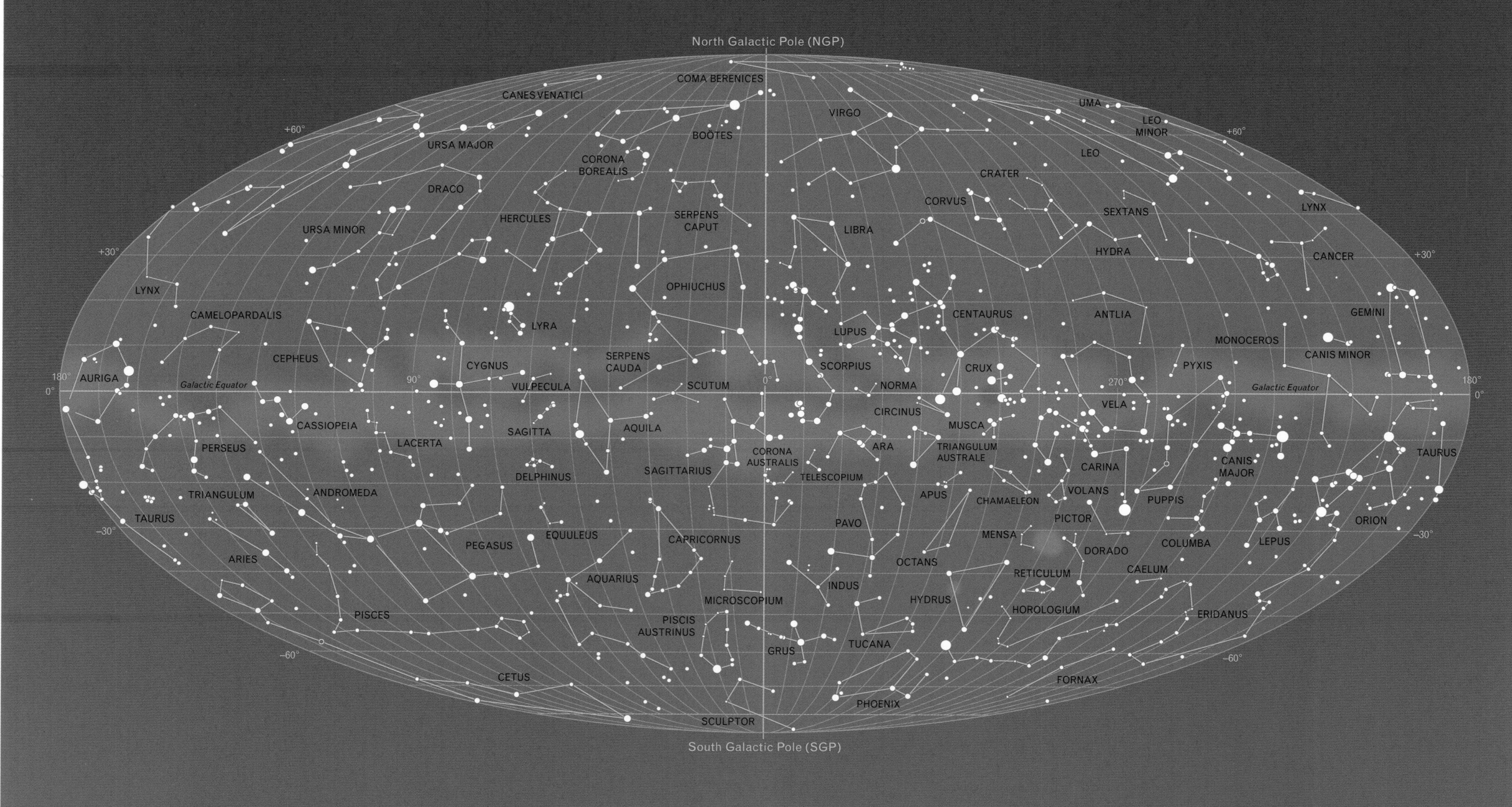
THE CONSTELLATIONS
North Galactic Pole (NGP)
South Galactic Pole (SGP)
+60°
+30°
0°
90°
180°
270°
–30°
–60°
Galactic Equator
COMA BERENICES
CANES VENATICI
URSA MAJOR
BOOTES
CORONA BOREALIS
DRACO
HERCULES
SERPENS CAPUT
URSA MINOR
VIRGO
LIBRA
CORVUS
CRATER
UMA
LEO MINOR
LEO
SEXTANS
HYDRA
LYNX
CANCER
LYNX
CAMELOPARDALIS
CEPHEUS
LYRA
OPHIUCHUS
CENTAURUS
LUPUS
ANTLIA
GEMINI
MONOCEROS
CANIS MINOR
AURIGA
CYGNUS
VULPECULA
SERPENS CAUDA
SCUTUM
SCORPIUS
NORMA
CRUX
PYXIS
VELA
CASSIOPEIA
PERSEUS
LACERTA
SAGITTA
AQUILA
CIRCINUS
MUSCA
CORONA AUSTRALIS
ARA
TRIANGULUM AUSTRALE
CARINA
CANIS MAJOR
TAURUS
TRIANGULUM
ANDROMEDA
DELPHINUS
SAGITTARIUS
TELESCOPIUM
APUS
CHAMAELEON
VOLANS
PUPPIS
TAURUS
PAVO
PICTOR
ORION
ARIES
PEGASUS
EQUULEUS
CAPRICORNUS
MENSA
DORADO
COLUMBA
LEPUS
OCTANS
CAELUM
AQUARIUS
INDUS
RETICULUM
MICROSCOPIUM
HYDRUS
HOROLOGIUM
PISCES
PISCIS AUSTRINUS
ERIDANUS
TUCANA
GRUS
CETUS
FORNAX
PHOENIX
SCULPTOR
MOLLWEIDE'S EQUAL-AREA PROJECTION
Galactic Coordinates

# DISTRIBUTION OF OPEN CLUSTERS

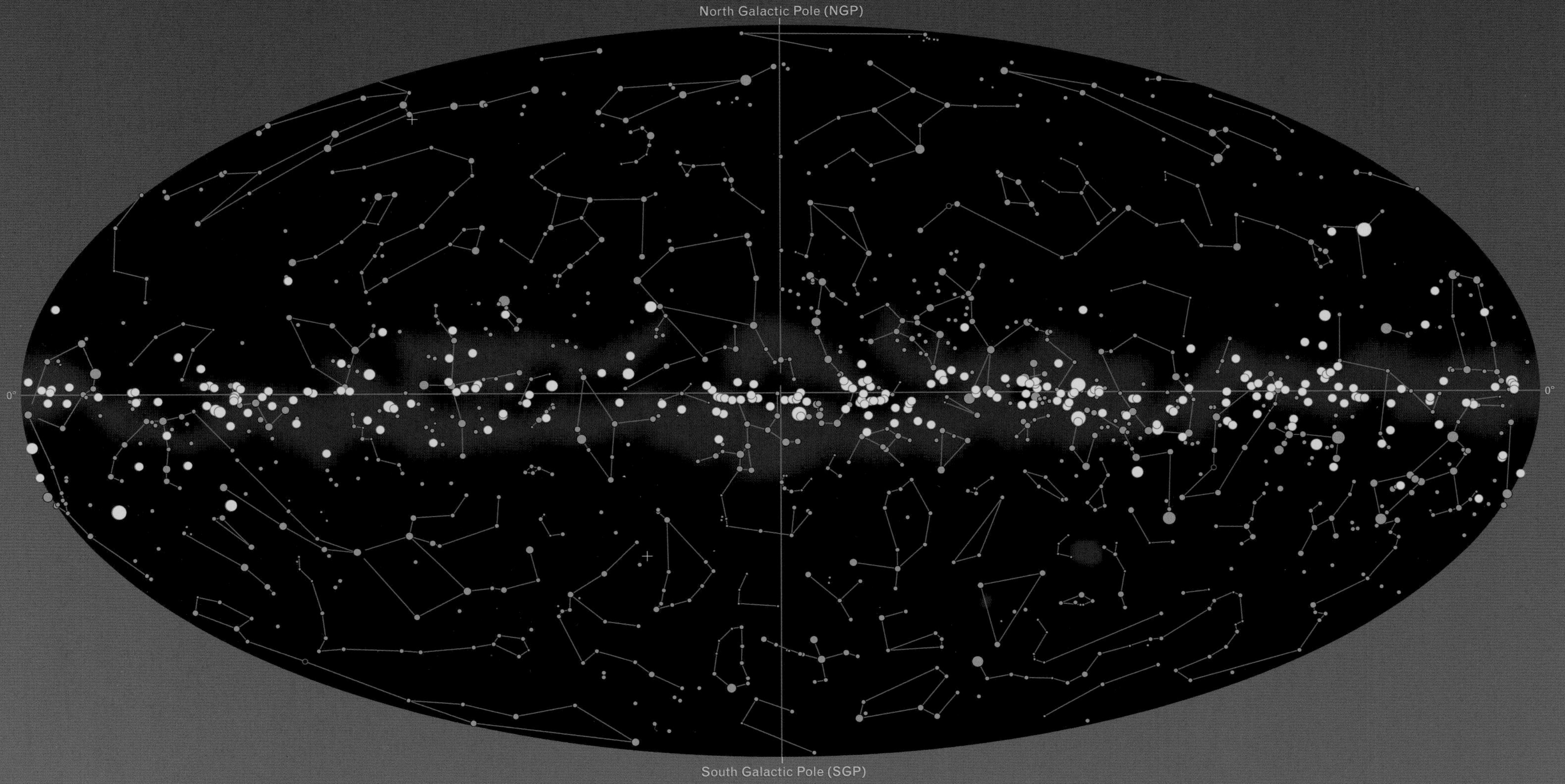

North Galactic Pole (NGP)

0°

0°

South Galactic Pole (SGP)

**MOLLWEIDE'S EQUAL-AREA PROJECTION**

Galactic Coordinates

# DISTRIBUTION OF GLOBULAR CLUSTERS

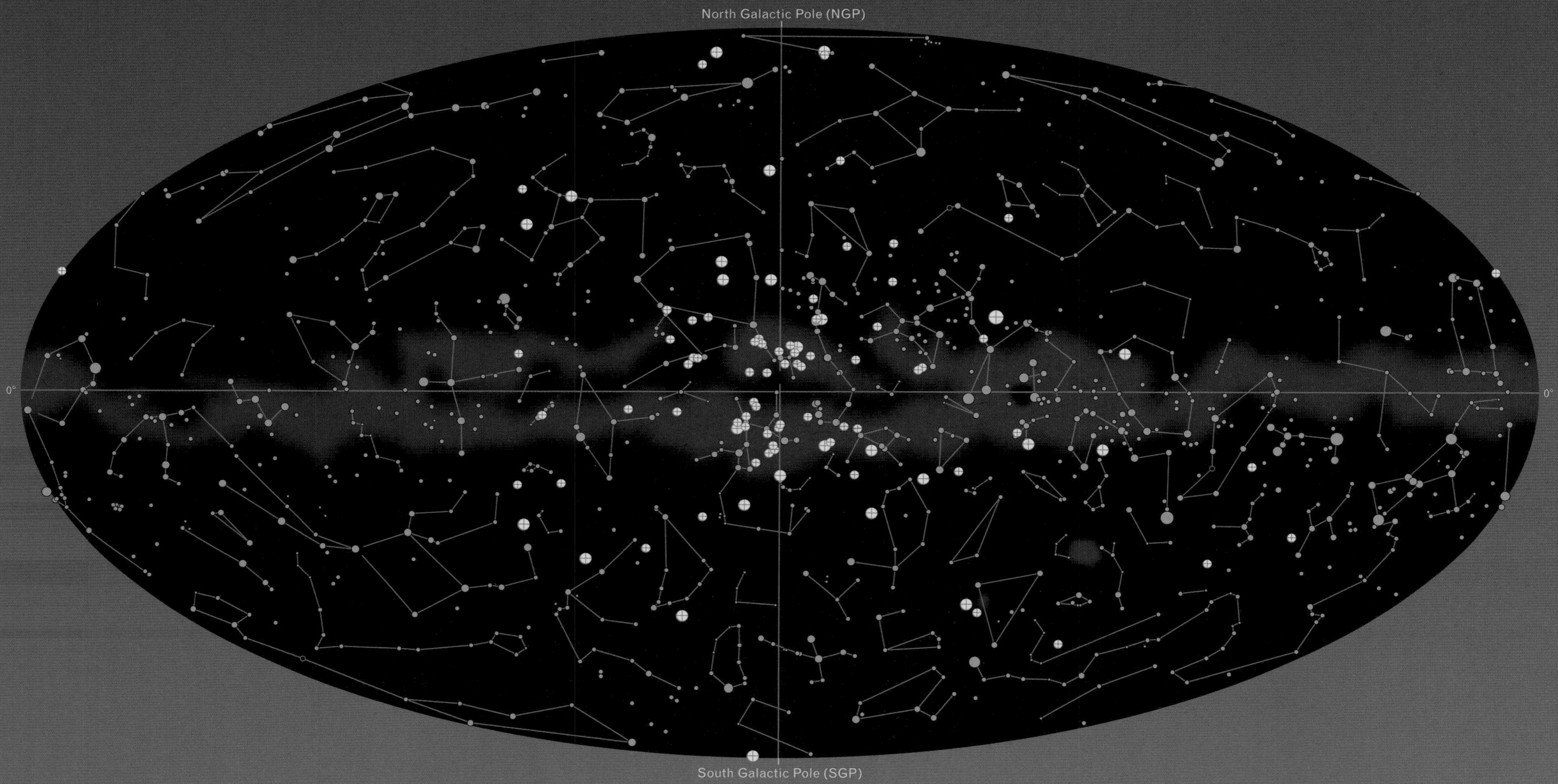

**MOLLWEIDE'S EQUAL-AREA PROJECTION**

Galactic Coordinates

# DISTRIBUTION OF DIFFUSE NEBULAE

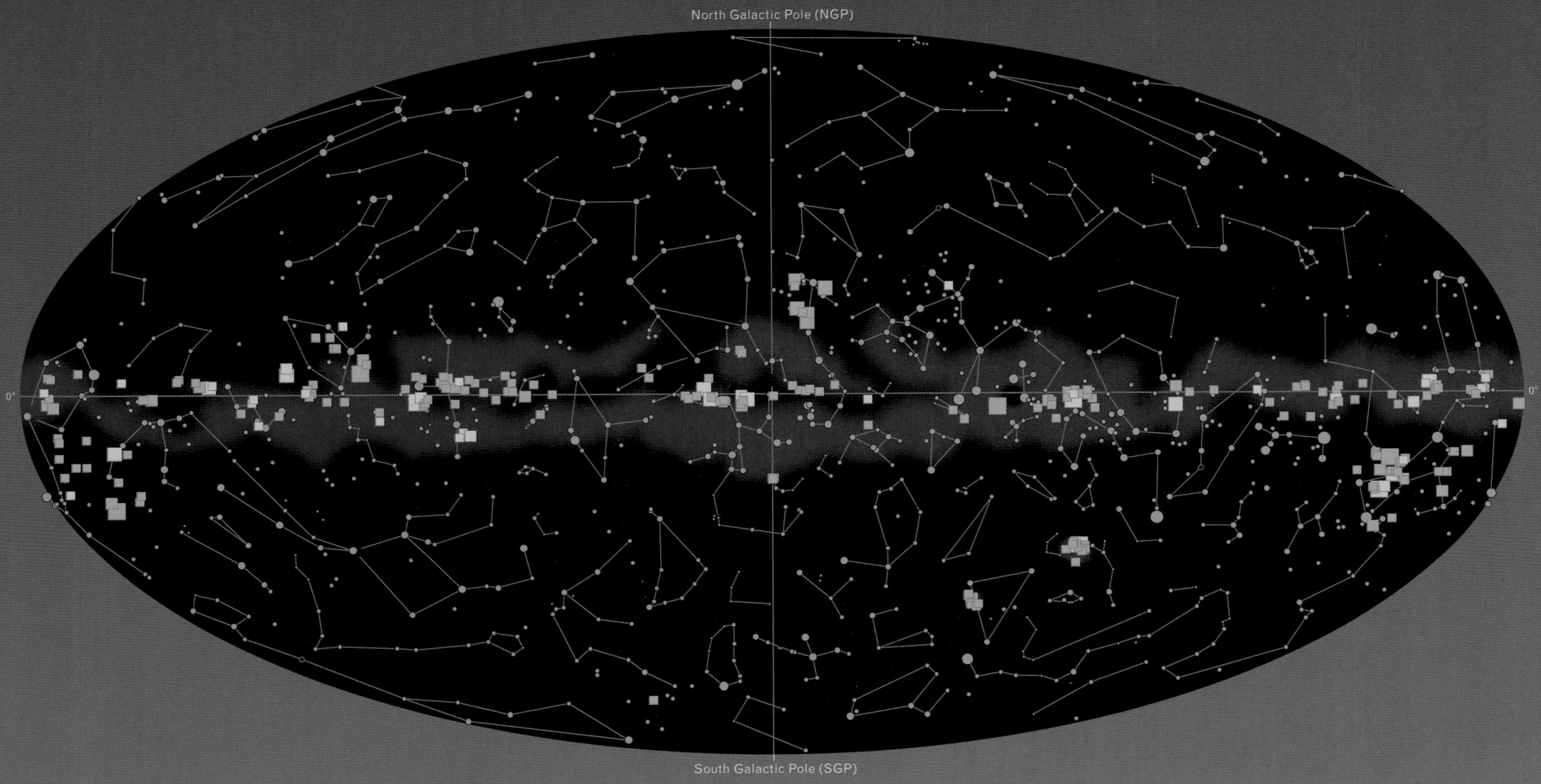

North Galactic Pole (NGP)

0°

0°

South Galactic Pole (SGP)

**MOLLWEIDE'S EQUAL-AREA PROJECTION**

Galactic Coordinates

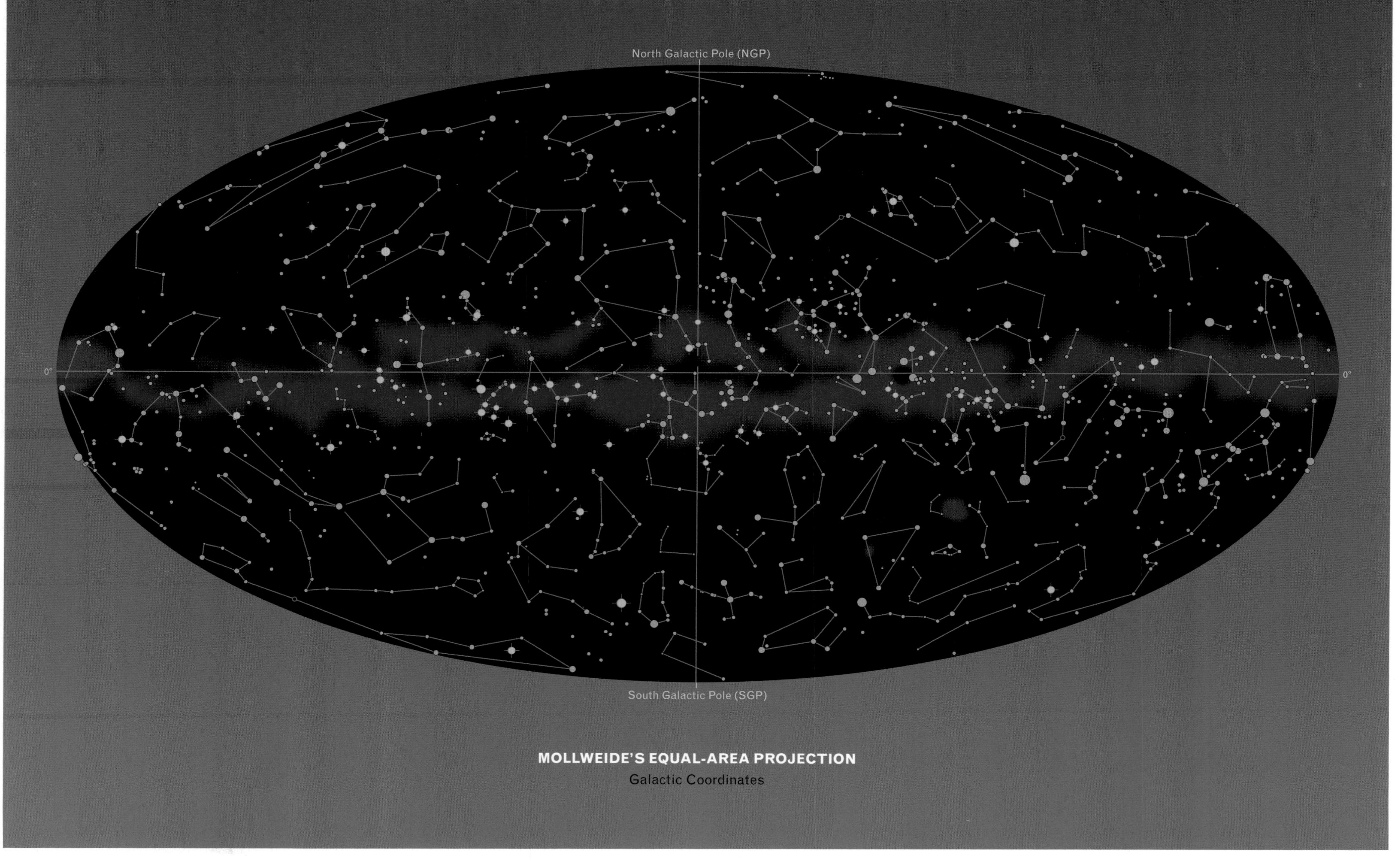
DISTRIBUTION OF PLANETARY NEBULAE
North Galactic Pole (NGP)
0°
0°
South Galactic Pole (SGP)
MOLLWEIDE'S EQUAL-AREA PROJECTION
Galactic Coordinates

# DISTRIBUTION OF GALAXIES

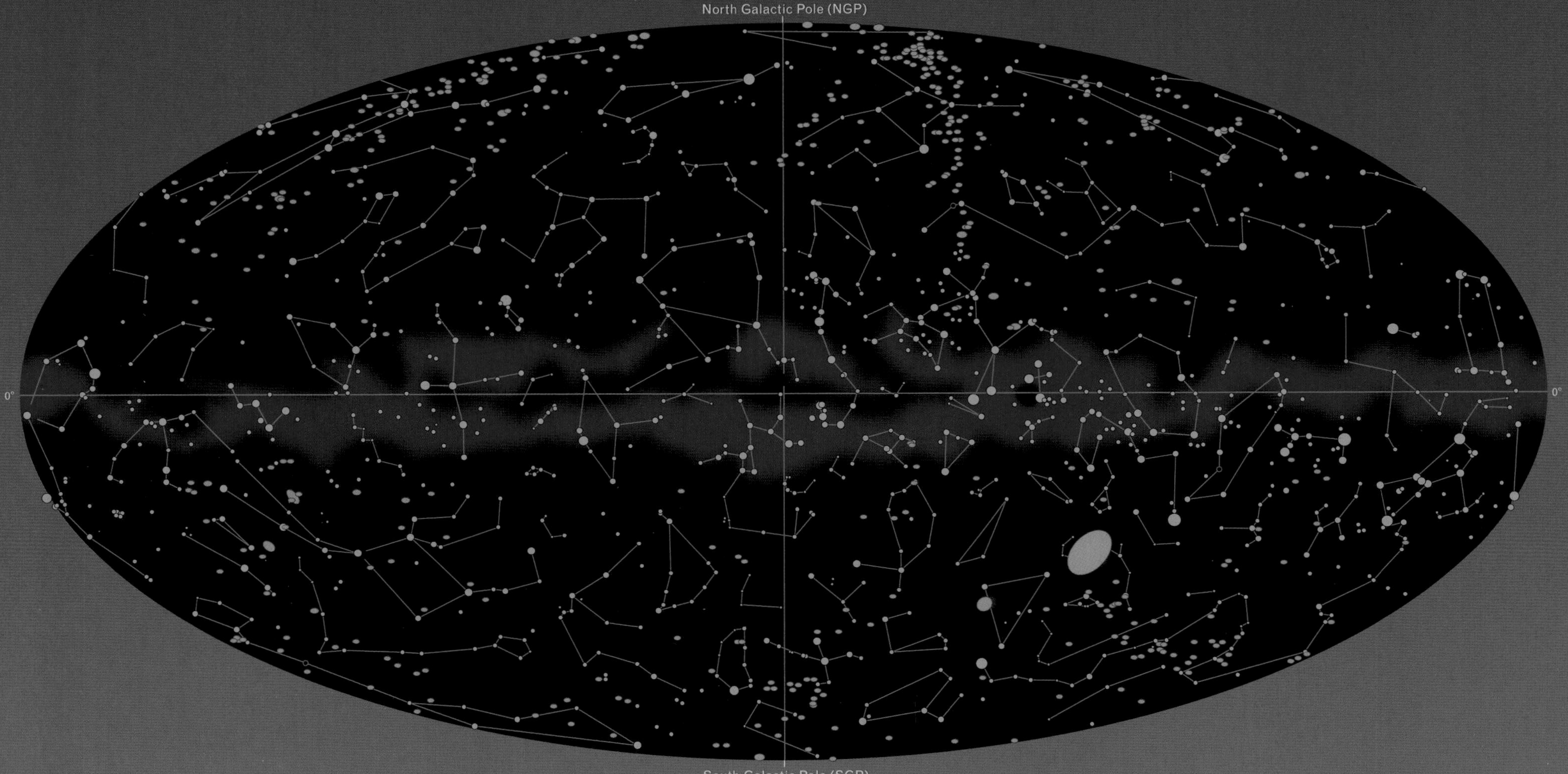

North Galactic Pole (NGP)

0°

0°

South Galactic Pole (SGP)

**MOLLWEIDE'S EQUAL-AREA PROJECTION**

Galactic Coordinates

# EXOPLANETS

*Exoplanets*, or *extrasolar planets*, are planets outside our solar system, orbiting other distant stars. For a long time it was not known if other stars had planet systems like our own Sun, or not. In 1995 a large planet was found orbiting the star 51 Pegasi. Soon more exoplanets were discovered and as of June 2010 there have been almost 500 confirmed discoveries.

For many people the discovery of extrasolar planets has intensified the hope of finding extra-terrestrial life, but most of the planets found so far are not solid Earth-like planets, but gas planets, like Jupiter, and not suitable to hold life as we know it.

Whether or not you believe in the possibility of extraterrestrial life, you might be interested to know what stars, plotted in this atlas, are found to have one or more planets. Table F provides you with this information. There are a total of 72 stars listed and the first six columns show the right ascension (RA) and declination (Dec) of the star, its name, its magnitude (Mag), the constellation (Con), and finally the maps where the star can be found. Note that not all star names are on the maps; the names in *italics* refer to stars that are plotted on the maps, but without labels. You can still locate them by using the RA and Dec information!

The last six columns provide information about the planets. The first planet found in a star's system receives the name of the star followed by a lower case b. The next planet is called c and so on. As you can see in the list, several stars are found to have more than one planet. The next three columns show information about the planet's orbit around the star; the period in days, the semi-major axis of the orbit in AU, or *Astronomical Unit* (149,598,000 kilometers, the average distance from the Earth to the Sun), and the eccentricity of the orbit.

The last two columns tell you in what year the planet was discovered and what its mass is, compared to the largest planet in our own solar system – Jupiter.

To keep up-to-date on the subject you might visit the website of the Planetary Society: http://www.planetary.org/exoplanets/list.php

**Table F** *List of exoplanets*

| RA h | RA m | Dec ° | Dec ′ | Star name | Mag | Con | Map numbers | Planet | Orbital period (d) | Semi-major axis (AU) | Eccentricity | Year of discovery | Mass (Jupiter = 1) |
|---|---|---|---|---|---|---|---|---|---|---|---|---|---|
| 00 | 06.3 | –49 | 05 | *HD 142* | 5.70 | Phe | 14 19 | b | 337.1 | 0.98 | 0.38 | 2001 | 1 |
| 00 | 18.7 | –08 | 03 | *HD 1461* | 6.46 | Cet | 8 13 | b | 5.77 | 0.063 | 0.14 | 2009 | 0.024 |
| 00 | 39.4 | +21 | 15 | 54 Piscium | 5.80 | Psc | 2 7 8 13 | b | 62.23 | 0.284 | 0.63 | 2003 | 0.2 |
| 00 | 44.7 | –65 | 39 | *HD 4308* | 6.54 | Tuc | 14 19 20 | b | 15.56 | 0.114 | 0 | 2005 | 0.047 |
| 01 | 38.8 | +41 | 24 | υ Andromedae | 4.09 | And | 2 7 | b | 4.617 | 0.059 | 0.029 | 1996 | 0.69 |
| | | | | | | | | c | 241.5 | 0.83 | 0.245 | 1999 | 1.98 |
| | | | | | | | | d | 1,275 | 2.51 | 0.245 | 1999 | 3.95 |
| 01 | 42.5 | –53 | 44 | q[1] Eridani | 5.52 | Eri | 14 15 19 20 | b | 1,040 | 2.1 | 0.18 | 2003 | 0.91 |
| 01 | 44.9 | +20 | 05 | 109 Piscium | 6.29 | Psc | 2 8 | b | 1,080 | 2.13 | 0.11 | 2000 | 6.12 |
| 01 | 54.9 | –67 | 39 | $\eta^2$ Hydri | 4.70 | Hyi | 14 20 | b | 711 | 1.93 | 0.4 | 2005 | 6.54 |
| 01 | 57.2 | –10 | 15 | *HD 11964* | 6.42 | Cet | 8 | b | 37.82 | 0.229 | 0.15 | 2005 | 0.11 |
| | | | | | | | | c | 2,110 | 3.34 | 0.06 | 2005 | 0.61 |
| 02 | 10.7 | –50 | 50 | *Gliese 86* | 6.17 | Eri | 15 19 20 | b | 15.77 | 0.11 | 0.046 | 2000 | 4.01 |
| 02 | 37.0 | +24 | 39 | 30 Arietis B | 7.10 | Ari | 2 8 | b | 335.1 | 0.995 | 0.289 | 2009 | 9.88 |
| 02 | 37.0 | –34 | 35 | $\lambda^2$ Fornacis | 5.78 | For | 14 15 | b | 17.24 | 0.14 | 0.2 | 2009 | 0.069 |
| 02 | 37.7 | –03 | 24 | 81 Ceti | 5.65 | Cet | 8 | b | 952.7 | 2.5 | 0.206 | 2008 | 5.3 |
| 02 | 42.6 | –50 | 49 | ι Horologii | 5.40 | Hor | 14 15 20 | b | 311.3 | 0.91 | 0.24 | 1999 | 1.94 |
| 03 | 12.8 | –01 | 12 | 94 Ceti | 5.07 | Cet | 8 | b | 454 | 1.3 | 0.2 | 2003 | 2 |
| 03 | 17.7 | +31 | 08 | *HD 20367* | 6.53 | Ari | 2 3 | b | 500 | 1.25 | 0.23 | 2002 | 1.07 |
| 03 | 32.9 | –09 | 27 | ε Eridani | 3.73 | Eri | 14 15 16 | b | 2,502 | 3.39 | 0.702 | 2000 | 1.55 |
| 04 | 16.5 | –59 | 18 | ε Reticuli | 4.44 | Ret | 14 15 20 | b | 423.8 | 1.18 | 0.07 | 2000 | 1.28 |
| 04 | 28.6 | +19 | 11 | ε Tauri | 3.53 | Tau | 2 3 8 9 | b | 594.9 | 1.93 | 0.151 | 2007 | 7.6 |
| 04 | 48.6 | –05 | 40 | *HD 30562* | 5.77 | Eri | 9 | b | 1,157 | 2.3 | 0.76 | 2009 | 1.29 |

**Table F** *List of exoplanets (continued)*

| RA h m | Dec ° ′ | Star name | Mag | Con | Map numbers | Planet | Orbital period (d) | Semi-major axis (AU) | Eccentricity | Year of discovery | Mass (Jupiter = 1) |
|---|---|---|---|---|---|---|---|---|---|---|---|
| 05 09.6 | +69 38 | *HD 32518* | 6.46 | Cam | 1 3 | b | 157.5 | 0.59 | 0.01 | 2009 | 3.04 |
| 05 22.6 | +79 14 | *HD 33564* | 5.05 | Cam | 1 | b | 388 | 1.1 | 0.34 | 2005 | 9.1 |
| 05 37.2 | −80 28 | π Mensae | 5.67 | Men | 20 | b | 2,064 | 3.29 | 0.62 | 2001 | 10.35 |
| 05 46.6 | +01 10 | *HD 38529* | 5.94 | Ori | 9 | b | 14.31 | 0.129 | 0.29 | 2000 | 0.78 |
| | | | | | | c | 2,174 | 3.68 | 0.36 | 2002 | 12.7 |
| 05 47.3 | −51 04 | β Pictoris | 3.86 | Pic | 14 15 16 | b | 6,000 | 8 | 0 | 2008 | 8 |
| 06 30.8 | +58 10 | 6 Lyncis | 5.20 | Lyn | 2 3 4 | b | 899 | 2.2 | 0.134 | 2008 | 2.4 |
| 06 37.8 | −32 20 | *HD 47536* | 5.27 | CMa | 15 | b | 430 | | 0.2 | 2003 | 5 |
| | | | | | | c | 2,500 | | 0 | 2007 | 7 |
| 07 00.3 | −05 22 | *HD 52265* | 6.30 | Mon | 9 | b | 119 | 0.49 | 0.29 | 2000 | 1.13 |
| 07 31.8 | +17 05 | *HD 59686* | 5.45 | Gem | 3 4 9 10 | b | 303 | 0.911 | 0 | 2003 | 5.25 |
| 07 34.1 | −22 18 | *HD 60532* | 4.45 | Pup | 9 10 15 16 | b | 201.89 | 0.77 | 0.278 | 2008 | 3.15 |
| | | | | | | c | 607.1 | 1.58 | 0.038 | 2008 | 7.46 |
| 07 45.3 | +28 02 | Pollux (α Gem) | 1.15 | Gem | 3 4 9 10 | b | 589.6 | 1.69 | 0.02 | 2006 | 2.9 |
| | | | | | | b | 8.7 | 0.0785 | 0.1 | 2006 | 0.033 |
| | | | | | | c | 31.6 | 0.186 | 0.13 | 2006 | 0.038 |
| 08 40.2 | +64 20 | 4 ($\pi^2$) Ursae Majoris | 5.79 | UMa | 1 3 4 | b | 269.3 | 0.87 | 0.432 | 2007 | 7.1 |
| 08 47.7 | −41 44 | *HD 75289* | 6.36 | Vel | 15 16 | b | 3.51 | 0.046 | 0.054 | 1999 | 0.42 |
| 08 52.6 | +28 20 | 55 ($\rho^1$) Cancri | 5.95 | Cnc | 3 4 | b | 14.65 | 0.115 | 0.014 | 1996 | 0.842 |
| | | | | | | c | 44.24 | 0.24 | 0.086 | 2002 | 0.169 |
| | | | | | | d | 5,218 | 5.77 | 0.025 | 2002 | 3.835 |
| | | | | | | e | 2.817 | 0.038 | 0.07 | 2004 | 0.034 |
| | | | | | | f | 260 | 0.781 | 0.2 | 2007 | 0.144 |
| 09 28.7 | +45 36 | *HD 81688* | 5.41 | UMa | 1 3 4 | b | 184 | 0.81 | 0 | 2008 | 2.7 |
| 10 20.0 | +19 51 | $\gamma^1$ Leonis | 2.61 | Leo | 4 10 | b | 428.5 | 1.19 | 0.144 | 2009 | 8.78 |
| 10 22.2 | +41 14 | *HD 89744* | 5.74 | UMa | 4 5 | b | 156.6 | 0.89 | 0.67 | 2000 | 7.99 |
| 10 59.5 | +40 26 | 47 Ursae Majoris | 5.10 | UMa | 4 5 | b | 1,083 | 2.11 | 0.049 | 1996 | 2.6 |
| | | | | | | c | 2,190 | 3.39 | 0.22 | 2001 | 0.46 |
| 12 05.3 | +76 54 | *HD 104985* | 5.79 | Cam | 1 | b | 198.2 | 0.78 | 0.03 | 2003 | 6.3 |
| 12 39.3 | −08 00 | χ Virginis | 4.66 | Vir | 10 11 | b | 835.5 | 2.14 | 0.462 | 2009 | 11.09 |
| 13 18.4 | −18 19 | 61 Virginis | 4.74 | Vir | 11 17 | b | 4.215 | 0.05 | 0.12 | 2009 | 0.016 |
| | | | | | | c | 38.02 | 0.2175 | 0.14 | 2009 | 0.0573 |
| | | | | | | d | 123 | 0.476 | 0.35 | 2009 | 0.072 |
| 13 28.4 | +13 47 | 70 Virginis | 5.00 | Vir | 11 | b | 116.7 | 0.48 | 0.4 | 1996 | 7.44 |
| 13 47.3 | +17 27 | τ Boötis | 4.50 | Boo | 5 11 | b | 3.314 | 0.046 | 0.018 | 1996 | 3.9 |
| 14 02.4 | −27 26 | *HD 122430* | 5.48 | Hya | 11 17 | b | 345 | 1.02 | 0.68 | 2003 | 3.71 |
| 15 13.5 | −25 19 | 23 Librae | 6.45 | Lib | 11 17 18 | b | 258.2 | 0.81 | 0.233 | 1999 | 1.59 |
| | | | | | | c | 5,000 | 5.8 | 0.12 | 2009 | 0.82 |
| 15 17.1 | +71 50 | 11 Ursae Minoris | 5.02 | UMi | 1 5 | b | 516.2 | 1.54 | 0.08 | 2009 | 10.5 |
| 15 24.9 | +58 58 | ι Draconis | 3.31 | Dra | 1 5 6 | b | 511.1 | 1.275 | 0.7124 | 2002 | 8.82 |
| 15 51.2 | +35 39 | κ Coronae Borealis | 4.79 | CrB | 5 6 | b | 1,191 | 2.7 | 0.19 | 2007 | 1.8 |
| 16 01.1 | +33 18 | ρ Coronae Borealis | 5.40 | CrB | 5 6 | b | 39.85 | 0.22 | 0.04 | 1997 | 1.04 |
| 16 24.0 | −39 12 | *HD 147513* | 5.37 | Sco | 17 18 | b | 540.4 | 1.26 | 0.52 | 2003 | 1 |
| 17 20.6 | −19 20 | *HD 156846* | 6.50 | Oph | 12 18 | b | 359.5 | 0.99 | 0.8472 | 2007 | 10.45 |
| 17 44.2 | −51 50 | μ Arae | 5.15 | Ara | 17 18 19 | b | 654.5 | 1.5 | 0.31 | 2000 | 1.67 |
| | | | | | | c | 2,986 | 4.17 | 0.57 | 2004 | 3.1 |
| | | | | | | d | 9.55 | 0.09 | 0 | 2004 | 0.044 |
| | | | | | | e | 310.6 | 0.921 | 0.0666 | 2006 | 0.521 |
| 18 10.5 | +54 17 | *HD 167042* | 5.95 | Dra | 5 6 7 | b | 416.1 | 1.3 | 0.03 | 2008 | 1.6 |
| 18 26.0 | +65 34 | 42 Draconis | 4.82 | Dra | 1 6 7 | b | 479.1 | 1.19 | 0.38 | 2009 | 3.88 |
| 18 27.8 | −29 49 | *HD 169830* | 5.91 | Sgr | 18 | b | 225.6 | 0.81 | 0.31 | 2000 | 2.88 |
| | | | | | | c | 2,102 | 3.6 | 0.33 | 2003 | 4.04 |
| 18 43.6 | +36 33 | *HD 173416* | 6.01 | Lyr | 6 7 | b | 323.6 | 1.16 | 0.21 | 2009 | 2.7 |
| 19 15.6 | −24 11 | *HD 179949* | 6.25 | Sgr | 12 18 19 | b | 3.093 | 0.045 | 0.022 | 2000 | 0.95 |
| 19 41.9 | +50 31 | 16 Cygni B | 6.20 | Cyg | 6 7 | b | 799.5 | 1.68 | 0.689 | 1996 | 1.68 |
| 19 54.3 | +08 28 | ξ Aquilae | 4.72 | Aql | 12 13 | b | 136.8 | 0.68 | 0 | 2008 | 2.8 |
| 20 03.6 | +29 54 | *HD 190360* | 5.71 | Cyg | 6 7 | b | 2,891 | 3.91 | 0.36 | 2003 | 1.502 |
| | | | | | | c | 17.1 | 0.128 | 0.01 | 2005 | 0.057 |
| 20 16.1 | +04 35 | *HD 192699* | 6.43 | Aql | 12 13 | b | 351.5 | 1.16 | 0.149 | 2007 | 2.5 |
| 20 39.9 | +11 15 | *HD 196885* | 6.42 | Del | 12 13 | b | 1,333 | 2.37 | 0.462 | 2007 | 2.58 |
| 20 58.4 | +10 51 | 18 Delphini | 5.52 | Del | 13 | b | 993.3 | 2.6 | 0.08 | 2008 | 10.3 |
| 22 11.9 | +16 02 | *HD 210702* | 5.95 | Peg | 13 | b | 341.1 | 1.17 | 0.152 | 2007 | 2 |
| 22 53.6 | −48 34 | $\tau^1$ Gruis | 6.03 | Gru | 14 19 | b | 1,443 | 2.7 | 0.34 | 2002 | 1.49 |
| 22 54.7 | −70 04 | ρ Indi | 6.05 | Ind | 19 20 | b | 1,294 | 2.7 | 0.34 | 2002 | 2.1 |
| 22 57.5 | +20 46 | 51 Pegasi | 5.49 | Peg | 7 13 | b | 4.231 | 0.052 | 0 | 1995 | 0.468 |
| 22 57.6 | −29 37 | Fomalhaut (α PsA) | 1.82 | PsA | 14 19 | b | 320,000 | 115 | 0.11 | 2008 | 3 |
| 22 58.3 | −02 24 | *HD 217107* | 6.16 | Psc | 13 | b | 7.127 | 0.073 | 1.132 | 1998 | 1.33 |
| | | | | | | c | 4,210 | 5.27 | 0.517 | 1998 | 2.49 |
| 23 07.5 | +21 08 | *HR 8799* | 5.99 | Peg | 7 13 | b | 170,000 | 68 | 0 | 2008 | 7 |
| | | | | | | c | 69,000 | 38 | 0 | 2008 | 10 |
| | | | | | | d | 36,500 | 24 | 0 | 2008 | 10 |
| 23 31.3 | +39 14 | 14 Andromedae | 5.22 | And | 2 7 | b | 185.8 | 0.83 | 0 | 2008 | 4.8 |
| 23 39.3 | +77 38 | γ Cephei | 3.22 | Cep | 1 | b | 902.9 | 2.044 | 0.115 | 2003 | 1.6 |

# SOURCES AND REFERENCES

## Catalogues

*The Hipparcos and Tycho Catalogues*
European Space Agency, ESTEC, Noordwijk,
The Netherlands, 1997

*Sky Catalogue 2000.0, Volume 1*, second edition
Alan Hirshfeld, Roger W. Sinnott and François Ochsenbein (eds.)
Sky Publishing Corporation, Cambridge, Massachusetts, and Cambridge University Press, Cambridge, England, 1991 (first edn 1982)

*Sky Catalogue 2000.0, Volume 2*
Alan Hirshfeld and Roger W. Sinnott (eds.)
Sky Publishing Corporation, Cambridge, Massachusetts, and Cambridge University Press, Cambridge, England, 1985

*NGC2000.0*
Roger W. Sinnott (ed.)
Sky Publishing Corporation, Cambridge, Massachusetts, and Cambridge University Press, Cambridge, England, 1988

*The Catalog of Exoplanets*
http://www.planetary.org/exoplanets/list.php
The Planetary Society,
Pasadena, California, 2010

The basic photographic Moon map is courtesy
*Mark Rosiek and the U.S. Geological Survey,*
Flagstaff, Arizona

## Atlases

*Millennium Star Atlas*
Roger W. Sinnott and Michael A.C. Perryman
Sky Publishing Corporation, Cambridge, Massachusetts,
European Space Agency, ESTEC, Noordwijk, The Netherlands, 1997

*Sky Atlas 2000.0,* second edition
Wil Tirion and Roger W. Sinnott
Sky Publishing Corporation, Cambridge, Massachusetts, 1998 (first edn 1981)
Cambridge University Press, Cambridge, England, 1998 (first edn 1981)

*The Cambridge Double Star Atlas*
James Mullaney and Wil Tirion
Cambridge University Press, Cambridge, England, 2009

*The Cambridge Atlas of Herschel Objects*
James Mullaney and Wil Tirion
Cambridge University Press, Cambridge, England, 2011